HOMEWORK HELPERS

Physics

Greg Curran

Homework Helpers: Physics
Edited by Kristen Parkes
Typeset by Eileen Dow Munson
Cover design by Lu Rossman/Digi Dog Design
Printed in the U.S.A. by Book-mart Press

To order this title, please call toll-free 1-800-CAREER-1 (NJ and Canada: 201-848-0310) to order using VISA or MasterCard, or for further information on books from Career Press.

The Career Press, Inc., 3 Tice Road, PO Box 687,
Franklin Lakes, NJ 07417
www.careerpress.com

Library of Congress Cataloging-in-Publication Data

Curran, Greg, 1966-
Homework helpers. Physics / by Greg Curran.
p. cm.
Includes index.
ISBN 1-56414-768-1 (pbk.)
1. Physics. I. Title.

QC23.2C85 2005
530--dc22

2004058278

Dedication

The year 2005 marks two important anniversaries. First, it marks the 100th anniversary of *annus mirabilis*, the year that the 26-year-old Albert Einstein made himself known to the scientific community with the publication of three scientific papers, one of which would earn him the Nobel Prize.

On a more personal note, 2005 also marks the 20th anniversary of the year I started dating the woman who was to become my wife. When deciding which of these important people I should dedicate this book to, it was no contest. My lovely wife, Rosemarie, has been with me since I was an 18-year-old slacker, dreamer, and freeloader. Without her constant love, support, and inspiration, I wouldn't have grown into the man I am today, a 38-year-old slacker, dreamer, and freeloader. As for Einstein, I never even met the man.

Happy 20 years, Rosemarie! This one is all for you.

Acknowledgments

Once again, I would like to thank Jessica Faust of BookEnds-Inc., who got the Homework Helper series off the ground. Thanks also go out to Eileen Dow Munson, Stacey Farkas, Kristen Parkes, Michael Pye, Michael Lewis, and the people at Career Press, for making this project possible.

I would also like to thank the following people, in no particular order:

John Haag, who, in addition to being a great friend, was the person who encouraged me to teach physics. It is easy to trace a path of causality back from the publication of this book to his influence on my career path.

Fr. Mickey Corcoran, who helped me a great deal when I started teaching physics.

Mike Curtin, who is like my good-luck charm. It seems like every time this guy gives me a laptop, I write a book on it.

Matt Distefano, an excellent teacher and the author of *Homework Helpers: Biology*, who is always there when I need to bounce ideas off someone.

Annette Ferranto, whose babysitting services allowed me to meet my deadline, once again.

Bernard Cornwell, who I don't know personally, but who dispelled the "myth" of "writer's block" in my mind, for all times.

My father and my brother, Peter F. Curran senior and junior, for chcking the text for errors.

To the entire Science department of Fordham Preparatory School, a dedicated and professional group of teachers, colleagues, and friends.

Once again, thank you all!

Contents

Preface 11

Introduction—Necessary Skills 13
- Use of Units 13
- Significant Digits 15
- Scientific Notation 17
- Sign Conventions 18
- Air Resistance 18
- Symbols and Formulas 19
- Algebra 20
- Trigonometry 22

Chapter 1—Kinematics 27
- Lesson 1–1: Distance and Displacement 27
- Lesson 1–2: Speed and Velocity 34
- Lesson 1–3: Acceleration 39
- Lesson 1–4: Constant Acceleration 43
- Lesson 1–5: Free Fall 47
- Lesson 1–6: Graphing Motion 53
- Lesson 1–7: Projectile Motion 61
- Chapter 1 Examination 70
- Answer Key 73

Chapter 2—Forces and the Laws of Motion 79

Lesson 2–1: Forces 79

Lesson 2–2: Newton's First Law of Motion 83

Lesson 2–3: Newton's Second Law of Motion 85

Lesson 2–4: Newton's Third Law of Motion 90

Lesson 2–5: Motion Along a Horizontal Surface 94

Lesson 2–6: Motion on an Inclined Surface 100

Chapter 2 Examination 107

Answer Key 109

Chapter 3—Work, Energy, Power, and Momentum 115

Lesson 3–1: Work 115

Lesson 3–2: Energy 119

Lesson 3–3: Conservation of Energy 125

Lesson 3–4: Power 129

Lesson 3–5: Linear Momentum 132

Lesson 3–6: Impulse 139

Chapter 3 Examination 141

Answer Key 143

Chapter 4—Rotational and Circular Motion 147

Lesson 4–1: Rotational Motion 147

Lesson 4–2: Torque 152

Lesson 4–3: Circular Motion 156

Lesson 4–4: Centripetal Force and Centripetal Acceleration 159

Lesson 4–5: Newton's Law of Gravity 162

Lesson 4–6: Kepler's Laws and the Motion of Satellites 165

Chapter 4 Examination 167

Answer Key 169

Chapter 5—Electric Charges, Forces, and Fields 173

Lesson 5–1: Electric Charges 173

Lesson 5–2: Electric Forces 176

Lesson 5–3: Methods for Charging Objects 180
Lesson 5–4: Electric Fields 184
Lesson 5–5: Electric Potential Energy 187
Lesson 5–6: Potential Difference 192
Lesson 5–7: Capacitance 195
Chapter 5 Examination 200
Answer Key 202

Chapter 6—Electric Current and Circuits 207
Lesson 6–1: Current 207
Lesson 6–2: Resistance 210
Lesson 6–3: Electric Power 213
Lesson 6–4: Circuits and Schematic Diagrams 217
Chapter 6 Examination 227
Answer Key 229

Chapter 7—Magnetism 233
Lesson 7–1: Magnets and Magnetic Fields 233
Lesson 7–2: Magnetic Fields Around Current-Carrying Wires 236
Lesson 7–3: Magnetic Field Strength 240
Lesson 7–4: Electromagnetic Induction 245
Chapter 7 Examination 250
Answer Key 252

Chapter 8—Waves and Light 255
Lesson 8–1: Types of Waves 255
Lesson 8–2: Properties and Characteristics of Waves 258
Lesson 8–3: Sound Waves 265
Lesson 8–4: Light 267
Lesson 8–5: Reflection and Mirrors 268
Lesson 8–6: Refraction and Lenses 275
Chapter 8 Examination 284
Answer Key 285

Chapter 9—Heat and Thermodynamics 291
Lesson 9–1: Heat and Temperature 291
Lesson 9–2: Heat Transfer 294
Lesson 9–3: Thermal Expansion 297
Lesson 9–4: The Gas Laws 300
Lesson 9–5: The Laws of Thermodynamics 306
Chapter 9 Examination 309
Answer Key 311

Chapter 10—Nuclear Physics 315
Lesson 10–1: Structure of the Atom 315
Lesson 10–2: Planck's Photons 318
Lesson 10–3: Binding Energy 321
Lesson 10–4: Nuclear Reactions 323
Chapter 10 Examination 327
Answer Key 328

Glossary 331

Index 341

About the Author 351

Preface

Welcome to *Homework Helpers: Physics*!

Physics. Is it a subject for everyone? Or is it only for the so-called "geek" or "nerd" who skips the school dance to work on his or her robot? Do you need to love your TI-83 calculator more than your family to really enjoy physics? Consider this: Physics is the study of the laws that govern the universe around us, and because we all live in this universe, aren't we all entitled to study these laws? Shouldn't we all have a chance to understand the mysteries of the world around us? Can't we all come to appreciate the implications of these laws in our own way?

It saddens me to think of the many people who avoid taking physics in high school because it has the reputation of being such a hard course. This reputation may be perpetuated by students who do poorly in the subject, after failing to put in the required effort. Some parents who have bad memories of their own experiences in physics class may pass this idea on to their children. This notion may also be reinforced by certain physics teachers who, believing that their subject isn't for everyone, discourage large enrollment in their classes.

I believe that physics is for everyone. Every student should take the opportunity to learn something about this fascinating branch of science. I recognize that everyone has different aptitudes, and physics won't come as easily to some as it will to others, but that doesn't change my opinion. Some students learn the subject quickly and painlessly, with minimal additional support. Other students will struggle with physics, but if they put in the required effort and they are given the additional support that they need, they can still do quite well and learn to enjoy the subject.

Homework Helpers: Physics has been written with the latter type of student in mind. My original intention with this book, as with *Homework Helpers: Chemistry* before it, is to serve as a sort of private tutor. I laid out the chapters and lessons in an order that should be similar to the order of your lessons in class. If you aren't quite getting everything in your physics class, and you want to read additional explanations and try additional problems to build your confidence in the subject, this book can be an invaluable resource. If you want to read up on a subject before your teacher lectures on it, but you find your textbook to hard to follow, give this book a try.

On the other hand, if you have already taken a course in physics and you feel like you could use a refresher, *Homework Helpers: Physics* can be of service to you as well. You can use this book to brush up on your high school physics before taking a course in college, or before taking the SAT II or AP physics examination.

The important thing is that you give the subject of physics a chance to earn a special place in your heart. Don't be intimidated by things that you have seen or heard about the subject in the past. With the right attitude and effort, you can master the subject, and with knowledge comes appreciation.

Good luck with your studies!

Introduction

Necessary Skills

Physics is a skill-based course. Students who begin their study of physics without the appropriate skill base must develop their skills quickly or risk struggling with the course. It is an unfortunate fact that some students fail to learn physics not because of the science, but because of the mathematics required in the course. The best way to keep the frustration level low is to keep the skill level high. In this section, I will quickly identify and review some of the skills and concepts that will help you succeed in physics. For the sake of space, this will represent a brief review, as I want to fill most of these pages with physics rather than basic math.

Use of Units

Like chemistry, physics is a quantitative subject, which uses numbers and units to represent physical quantities. Each number must be recorded with a unit, otherwise its meaning can be misunderstood. For example, if you read the recorded speed of a car was 45, with no units, you could not be sure of how fast the car was traveling. You might assume that it meant 45 miles per hour, but it could just as easily have been 45 meters per second or 45 kilometers per hour. To avoid confusion, we would record a speed of 45 meters per second as 45 m/s. The sooner that you can get in the habit of recording units with every number, the better. This habit will have a positive effect on your exam scores.

We will be using the International System of Measurements (or SI), which you may be familiar with from earlier science courses. The SI limits the number of base units used in measurements to the following seven:

SI Base Units

Property	Unit	Symbol
Length	Meter	m
Time	Seconds	s
Mass	Kilogram	kg
Temperature	Kelvin	K
Electric Current	Ampere	A
Amount of Substance	Mole	mol
Luminous Intensity	Candela	cd

As you study physics, you will encounter the first five properties and units fairly often. When you encounter other units, such as miles or minutes, you may need to convert to SI units in order to solve a problem correctly. You will also encounter a number of derived units. **Derived units** are some combination of base units, which are often given a special name and symbol. For example, the newton (N) is a common derived unit for force. One newton is equivalent to 1 $kg \cdot m/s^2$. When doing calculations with derived units, you may need to make conversions to like units.

Examples of SI Derived Units

Property	Unit	Symbol	Equivalent Base Units
Force	Newton	N	$kg \cdot m/s^2$
Energy	Joules	J	$kg \cdot m^2/s^2$
Electric Charge	Coulomb	C	$A \cdot s$
Frequency	Hertz	Hz	s^{-1}

For example, it wouldn't be appropriate to divide a value in newtons by another value in minutes. You would change the minutes to seconds before doing the calculation.

Significant Digits

Not only should each number be recorded with units, each number should be recorded to the correct number of significant digits. You should find out, as early as possible, if your instructor will require you to round each answer to the correct number of significant digits. Students who never learn to round correctly may end up losing a point on each calculation on every quiz or exam they take each year, and all those points add up! Even if your instructor doesn't require significant digits, most standardized exams do. It is certainly in a student's best interest to learn to round properly as early as possible.

There are two important parts to working with significant digits. First, you need to be able to identify the number of significant digits shown in a number. Second, you need to be able to round answers correctly, based on the number of significant digits in the original problem. I will present a quick review of both of these topics now, but for a much more detailed explanation you can read *Homework Helpers: Chemistry*, also from Career Press.

You determine the number of significant digits shown in a number by following these rules:

1. Any nonzero digits are always considered significant.
2. Zeros that are found between two significant digits are significant.
3. Zeros that are found to the right of *both* a decimal and another significant digit are significant.
4. Zeros that appear solely as placeholder are not considered significant.

In addition to these rules, it is important to note that **counting numbers** are considered to have an infinite number of significant digits. Counting numbers are things that you actually count, rather than measure.

When dealing with a number that is written in scientific notation, you only analyze the coefficient when determining the number of significant digits present.

Examples for Determining the Number of Significant Digits

- **453** has three significant digits, because of rule one: each digit is a nonzero digit.
- **3007** has four significant digits because of rules one and two: The 3 and the 7 are nonzero digits, and both zeros are found in between two other significant digits.
- **35.00** has four significant digits because of rules one and three. The 3 and the 5 are nonzero digits, and both zeros are found to the right of both a decimal and a significant digit.
- **0.0008** has one significant digit because of rules one and four: The 8 is a nonzero digit and is significant. The four zeros shown in the number are just placeholders. Remember, rule three says that the zeros must be to the right of *both* a decimal point and a significant digit to be considered significant.
- **400** has only one significant digit because of rules one and four: The 4 is a nonzero digit, making it significant. The two zeros are not covered by rules two or three, so they are merely placeholders.
- $\mathbf{4.00 \times 10^3}$ has three significant digits because of rules one and three: The 4 is a nonzero digit and the two zeros are shown to the right of *both* a decimal place and a significant digit.

When you do calculations based on measurements, you must round your answer to the proper number of significant digits, based on the following rules:

1. For multiplication and division, your answer must show the same number of significant digits as the measurement in the calculation with the least number of significant digits. For example,

   $3.40 \text{ cm} \times 12.61 \text{ cm} \times 18.25 \text{ cm} = 782.4505 \text{ cm}^3$ before rounding.

   We can't report an answer with seven significant digits if the measurement with the least number of significant digits in our calculation, 3.40 cm, shows only three significant digits. We must round our answer to three significant digits, giving us a rounded answer of $\mathbf{782 \text{ cm}^3}$.

2. For addition and subtraction, your answer must show the same number of decimal places as the number in the calculation with the least number of decimal places. For example,

   22.530 m/s – 8.07 m/s = 14.46 m/s.

   In this example, our answer would be correct as shown. The measurement with the least number of decimal places, 8.07 m/s, is reported to the hundredths place. Our answer must also be reported to the hundredth place as 14.46 m/s.

There will be times when you see answers, in this book and others, that are written in scientific notation, and you might not be sure why. Often, answers are written in scientific notation just to express the answer in the correct number of significant digits. Look at the following example.

$$8.0 \text{ m} \times 50.0 \text{ m} = 400 \text{ m}^2$$

The measurement in this calculation with the least number of significant digits, 8.0 m, shows two significant digits. Our answer only shows one significant digit. We can write this answer in scientific notation just for the purpose of showing the same value with two significant digits. Our answer would be expressed as $\mathbf{4.0 \times 10^2\ m^2}$.

There is another method for showing the number 400 with two significant digits. You might see it written with a line over the first zero, as in $4\bar{0}0$. The line indicates that the zero, which would not be considered significant under the rules we covered, is indeed significant.

A third method for showing additional significant digits involves using a decimal at the right of a significant digit of zero. For example, if you wanted to show the number 700 with three significant digits, you could put a decimal point to the right of the second zero, as in 700., or you could use either of the other two methods described here. For this book, we will be using decimals and/or the scientific notation method.

Scientific Notation

Physics is similar to chemistry, astronomy, and other sciences in that it deals with very large and very small numbers. To avoid having to write and work with long numbers, scientific notation is often employed. A number that is written in scientific notation consists of three parts: the coefficient, base, and exponent.

If the number is written properly, the coefficient will be greater than or equal to one, but less than 10. The base will be 10, and the exponent will be a number that indicates the number of places that the decimal in the coefficient needs to be moved. Examples of some numbers in scientific notation and the values they represent are shown here:

Examples of Numbers in Both Scientific and Standard Notation

Scientific Notation	Standard Notation
2.75×10^4	27 500
6.02×10^{23}	602 000 000 000 000 000 000 000
4×10^{-7}	0.0000004
8.5×10^0	8.5

Sign Conventions

Many of the calculations that you perform in physics will represent the motion of an object in one- or two-dimensional space. Other calculations will represent a transfer of energy from one system to another. When dealing with these calculations, it is easier to keep track of the direction of a change if we assign positive and negative signs to our values. For example, we can assign a positive sign to the heat that is gained by a body of water, and a negative sign to the heat that is lost by the body of water. Or we could assign a positive sign to the velocity of a baseball as it moves upwards, and a negative sign to the motion of the baseball as it moves downward.

The important thing to remember about these sign conventions is that they are arbitrary. One text may refer to acceleration in the downward direction as negative, while another refers to the same acceleration as positive. Depending on how you do your calculations, you may get the answer with the same direction, but the opposite sign. The reason for this may be as simple as you assigning a positive to a direction that the text is assigning a negative to. When working on calculations in physics, pay careful attention to signs.

Air Resistance

When an object moves around, on, or near Earth, it pushes past air molecules. For example, when you walk down a hallway, you displace air molecules as you move forward. These air molecules offer a certain amount of resistance, but under ordinary conditions, you don't feel it. If you have ever walked into a strong wind, you have experienced significant air resistance.

For a physics student, air resistance can be a tricky thing because it is so variable. Atmospheric conditions, such as air pressure, change from day to day and location to location. Rather than trying to take unpredictable air resistance into account when doing calculations, you will often be told to "ignore air resistance" or even to "ignore friction" in general. Some students feel that this omission makes the calculations that they are doing invalid. This is not the case. The calculations that you will learn do a good job of describing the world around you. The air resistance that you are ignoring is not always significant, and most of the motion that takes place in our universe takes place in areas of space where there would be no significant air resistance. More importantly, the air resistance on Earth is so variable that it is not possible to take it into account in every situation.

For your part, you should assume that the calculations that you see performed in class, and in this book, don't take air resistance into account unless otherwise noted. Don't let this make you feel that the problems that you carry out in class are "fake" or inaccurate. As you will learn from your lab experiences, these calculations will allow you to predict the motion of real objects to a great degree of accuracy.

Symbols and Formulas

Mathematics is the language of physics, and you can't study physics without encountering numerous symbols and formulas. Some symbols are used universally, and will likely appear the same in all texts. For example, acceleration always seems to be represented by a lowercase a. What may confuse you in the beginning is that not all quantities will be represented by the same symbol in every text or every class. Different teachers and different texts will often use different symbols to represent the same quantity. For example, some texts will use the symbol d to represent displacement, while others will use Δx or Δp or Δs to represent the same quantity. As a result of the use of different symbols to represent the same quantities, formulas using these symbols may appear quite different, even if they mean the same thing! For example, the following set of formulas show the same relationships, yet they appear quite different to a novice physics student.

Examples of Formulas That Look Different, Yet Represent the Same Calculation

$$v^2 = v_0^{\ 2} + 2ax$$

$$v^2 = v_0^{\ 2} + 2a(s - s_0)$$

$$v_f^{\ 2} = v_i^{\ 2} + 2a(d - d_0)$$

$$v_f = \sqrt{v_i^{\ 2} + 2a\Delta x}$$

If all of these formulas look so different, is there one that is the "correct" formula? That's like asking which of several sketches of a person is the "correct" sketch. All of the formulas shown actually represent the same thing, and an experienced physics or math student will probably see that right away. Yet beginning physics students usually only become familiar with the "look" of the formulas and symbols that they see in class or in their particular text. A problem might arise when they pick up a different book, or if they take a standardized test and are given a formula sheet with formulas that they fail to recognize.

The solution to the problem is quite simple. Don't grow too attached to a particular symbol or formula. Think of the letters or symbols that are used to represent physical quantities as arbitrary. Instead of trying to memorize how a particular formula looks, try to understand and recall the relationships between the physical quantities involved. For example, if you remember the relationship between speed, distance, and time,

$$\text{speed} = \frac{\text{distance}}{\text{time}},$$

it won't matter to you if distance is represented by d or x or Δp, because you can still recognize or derive a proper formula.

Algebra

There is probably no more significant barrier to learning introductory physics than a weakness in algebra. Students who are unable to isolate each of the possible unknown quantities in the many formulas they encounter in physics tend to run into trouble in the first couple weeks of class. If your study of physics uncovers a weakness in algebra, it is best to take care of that problem first, so that you don't develop a negative attitude toward the science. Fortunately, the algebraic skills necessary for the study of basic physics are relatively simple, and a brief review may be all that you need.

Let's look at an example of a physics formula, and go through the isolation of each of the possible unknown quantities. If you want to test yourself, write the following equation on a piece of paper and try to isolate each of the variables, one at a time. Then check back here to compare your answer to the work shown following the equation.

$$d = \frac{1}{2} g \Delta t^2$$

The formula is already shown with d isolated. Let's go over methods for isolating g and Δt.

Solving for g

Original formula: $d = \frac{1}{2} g \Delta t^2$

Multiply both sides by 2: $2 \times d = \cancel{2} \times \frac{1}{\cancel{2}} g \Delta t^2$

$2 \times \frac{1}{2} = 1$ so, we get: $2d = g \Delta t^2$

Now, divide both sides by Δt^2: $\frac{2d}{\Delta t^2} = \frac{g \cancel{\Delta t^2}}{\cancel{\Delta t^2}}$

This leaves us with: $g = \frac{2d}{\Delta t^2}$

Now, we will go back to the original formula and solve for Δt.

Solving for Δt

Original formula: $d = \frac{1}{2} g \Delta t^2$

Multiply both sides by 2: $2 \times d = \cancel{2} \times \frac{1}{\cancel{2}} g \Delta t^2$

$2 \times \frac{1}{2} = 1$ so, we get: $2d = g \Delta t^2$

Now, divide both sides by g: $\frac{2d}{g} = \frac{g\Delta t^2}{g}$

This leaves us with: $\Delta t^2 = \frac{2d}{g}$

Now, take the square root of both sides: $\sqrt{\Delta t^2} = \sqrt{\frac{2d}{g}}$

Which leaves us with the final equation: $\Delta t = \sqrt{\frac{2d}{g}}$

Of course, you don't have to follow the steps shown here in this exact order. The important thing is that the final equations look the same. If you were able to successfully isolate the unknown quantities in the preceding formula, then you probably possess the algebraic skills required to succeed in physics. If, however, you need more help, I would recommend that you purchase an algebra review book.

Trigonometry

Trigonometry is the branch of mathematics concerned with studying the properties of right triangles. At first, it might not be clear to you how this may pertain to the study of physics, so I will give you an example. Suppose a person was pushing on the handle of a shopping cart, exerting a force of 20.0 newtons (N) at an angle of 30.0° below the horizontal, represented by the black arrow in Figure 1.1. In order to calculate the acceleration of the shopping cart, you want to know how much of the applied force (F), or push, is being exerted "downwards" and how much of the force is being exerted "forwards."

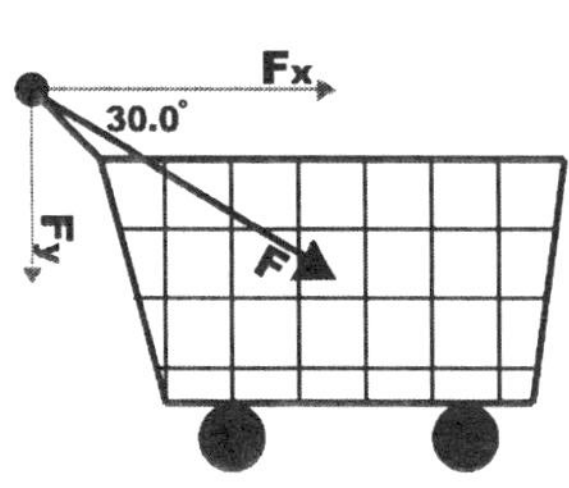

Figure 0.1

Notice that the downwards component of the force (F_y) is perpendicular to the forwards, or horizontal, component of the force (F_x). This means that the components are at a right angle to each other. We can sketch the problem that we are trying to solve as a right triangle, using the original force of 20.0 N as the hypotenuse, and the horizontal and vertical components as the other sides. Furthermore, we can add the angles that we know. Because all of the angles in a triangle must add up to 180°, we can determine the third angle as well, but I will leave it out of Figure 0.2 to avoid confusion.

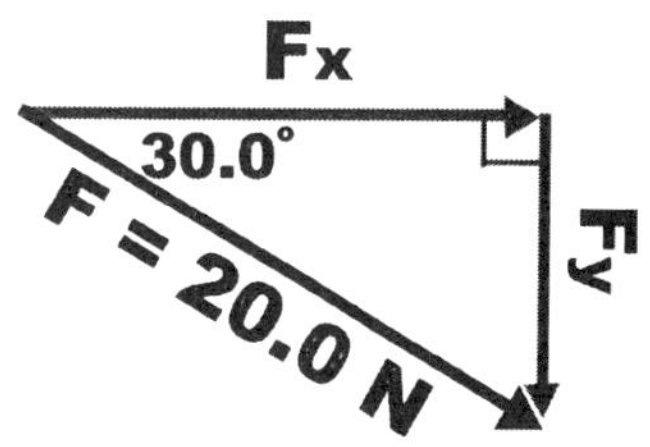

Figure 0.2

We can solve for the missing sides of our triangle using some of the following trigonometry functions.

Functions for Finding a Missing Side

$$\text{Sine (sin): } \sin\theta = \frac{\text{opposite side}}{\text{hypotenuse}}$$

$$\text{Cosine (cos): } \cos\theta = \frac{\text{adjacent side}}{\text{hypotenuse}}$$

$$\text{Tangent (tan): } \tan\theta = \frac{\text{opposite side}}{\text{adjacent side}}$$

The hypotenuse of our triangle is the original force of 20.0 N. Working with our original 30.0° angle, we can see that the vertical component of the force (F_y) represents the opposite side. Knowing this angle and the hypotenuse, we can solve for the vertical component (opposite side), as shown here:

$$\text{Original formula: } \sin\theta = \frac{\text{opposite side}}{\text{hypotenuse}}$$

Isolating the opposite side:

$$\text{hypotenuse} \times \sin\theta = \frac{\text{opposite side}}{\text{hypotenuse}} \times \text{hypotenuse}$$

We get:

$$\text{opposite side} = \text{hypotenuse} \times \sin\theta = (20.0\ \text{N})(\sin 30.0^\circ) = \mathbf{10.0\ N}$$

So, the vertical component of the original force is **10.0 N**.

We can also find the horizontal component of the original force, keeping in mind that it represents the adjacent side to our original angle of 30.0°.

Original formula: $\cos\theta = \dfrac{\text{adjacent side}}{\text{hypotenuse}}$

Isolating the adjacent side:

$$\text{hypotenuse} \times \cos\theta = \frac{\text{adjacent side}}{\text{hypotenuse}} \times \text{hypotenuse}$$

We get:

adjacent side = hypotenuse × cos θ = (20.0 N)(cos 30.0°) = 17.3 N

So, the horizontal component of the original force is **17.3 N.**

Alternatively, we could have found the last missing side using the Pythagorean theorem:

$a^2 + b^2 = c^2$

$$a = \sqrt{c^2 - b^2} = \sqrt{(20.0\text{ N})^2 - (10.0\text{ N})^2} = \mathbf{17.3\text{ N}}$$

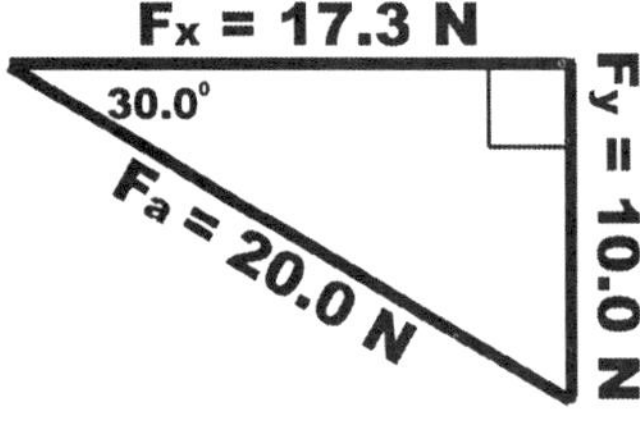

Figure 0.3

At other times, you will want to use trigonometry functions for finding the missing angle in a triangle. Let's suppose there is a river where the water flows with a velocity of 8.00 m/s (meters per second) east. A man crosses the river in a boat with a velocity of 20.0 m/s north, relative to the water. The water pushes the boat east during the crossing, so that to an observer on the shore, the boat seems to head northeast as it crosses. How would we find both the magnitude and direction of the resultant velocity?

The velocity of the boat (v_b) and the water (v_w) are perpendicular to each other, so finding the magnitude of the resultant velocity isn't hard.

We can turn the sketch into a right triangle and use the Pythagorean theorem again.

$$c^2 = a^2 + b^2$$

$$c = \sqrt{a^2 + b^2} = \sqrt{(8.0\,\text{m/s})^2 + (20.0\,\text{m/s})^2} = \mathbf{21.5\,m/s}$$

To find the angle between the intended path and the actual path of the boat, we make use of one of the following trigonometry functions.

Functions for Finding a Missing Angle

Inverse sine ($\sin^{-1}$): $\theta = \sin^{-1}\left(\frac{\text{opposite side}}{\text{hypotenuse}}\right)$

Inverse cosine ($\cos^{-1}$): $\theta = \cos^{-1}\left(\frac{\text{adjacent side}}{\text{hypotenuse}}\right)$

Inverse tangent ($\tan^{-1}$): $\theta = \tan^{-1}\left(\frac{\text{opposite side}}{\text{adjacent side}}\right)$

—

It makes sense to use the inverse tangent for this problem, so that we can make use of the numbers that came with the actual problem. Personally, I would avoid using a function involving the hypotenuse in this particular case, because if we made an error finding that value, the error would carry over to our answer for the angle. Solving for the angle we find:

$$\theta = \tan^{-1}\left(\frac{\text{opposite side}}{\text{adjacent side}}\right) = \tan^{-1}\left(\frac{20.0\,\cancel{\text{m/s}}}{8.00\,\cancel{\text{m/s}}}\right) = \mathbf{68.2^{\circ}}$$

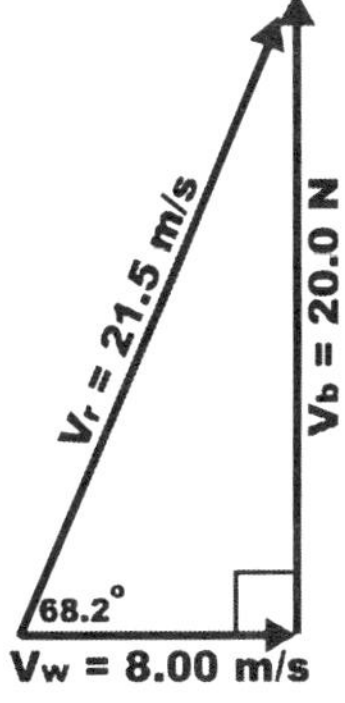

Figure 0.4

So, the resultant velocity (v_r) of the boat crossing the river is **21.5 m/s at an angle of 68.2° north of east.**

1 2 3 4 5 6 7 8 9 10

Kinematics

We begin our study of physics with the branch called **mechanics**. Mechanics is the study of the motion of objects. **Classical,** or **Newtonian, mechanics** deals with the motion of macroscopic, or relatively large, objects. **Quantum mechanics** deals with the motion of microscopic particles. **Kinematics** is the study of the motion of objects without regard for the forces that influence the objects. In this chapter, we will concern ourselves solely with analyzing the motion of objects, not the forces that can change motion. We will begin to study forces in Chapter 2. The formulas for kinematics will allow you to analyze the motion of objects, such as baseballs and cars, which you encounter in your day-to-day life. As you learn the formulas, you might want to look for opportunities to apply them to real life. They will not be hard to find.

Lesson 1–1: Distance and Displacement

When studying any type of motion, perhaps the first thing that a person is likely to notice is a change in the object's position. If we looked out a window and noticed a squirrel, we would have an opportunity to observe its motion. Of course, even a "stationary" squirrel is in constant motion from some point of view. The squirrel is on a planet that revolves and rotates in space. Furthermore, even when it appears to be standing still it may move its arms or tail, but in order to exhibit the type of motion that we are concerned with here, it would need to change its position relative to some "fixed" object, or reference point.

When we first noticed the squirrel, it might have been a certain distance away from a tree. Over time, it may move closer or farther away from the tree. Such a movement is referred to as a **change in position** and will be represented in this book by the symbol ΔX. Let's suppose we first observed the squirrel when it was 1.0 m away from the tree (our reference point). After a burst of speed in a straight line, the squirrel stops at a point that is 2.5 m away from the same tree. The change in the squirrel's position (**ΔX**) would be

$$\Delta X = X_{final} - X_{initial} = 2.5\text{ m} - 1.0\text{ m} = 1.5\text{ m}.$$

We might also call this quantity the **distance** that the squirrel traveled, so distance represents how far an object moves. In physics we are often concerned with a similar quantity called **displacement**, which we will represent with the symbol d. Displacement represents the change in an object's position in a particular direction. If we noted that the squirrel moved 1.5 m away from the tree in a direction of 25° east of south, we would be talking about the displacement of the squirrel. If we are comparing the distance and displacement of the squirrel in this example, it would be correct to describe the displacement as "the distance that the squirrel traveled in a particular direction." However, it is much better to think of displacement as the change in an object's position in a particular direction. The real distinction between distance and displacement will be made more apparent when we watch the squirrel for a bit longer.

Suppose we observed the squirrel for several more minutes in which it traveled a total of another 4.5 m, ultimately ending up in the same location (1.0 m away from the tree) as when we first observed it. Now, in the time that we had been observing it, the squirrel would have covered a total distance of (1.5 m + 4.5 m) 6.0 m. What would be the overall *displacement* of the squirrel in this amount of time? Zero! Zero? Yes, zero. Because the squirrel ended up on its original starting point, the overall change in the squirrel's position, measured from the tree, or origin, is

$$d = \Delta X = X_{final} - X_{initial} = 1.0\text{ m} - 1.0\text{ m} = 0.0\text{ m}.$$

Distance, and other quantities that can be represented completely with numbers and units, are called **scalars** or **scalar quantities**. Displacement, on the other hand, is the first of several **vector quantities** that you will encounter in physics. A vector quantity, or just vector for short, must be represented by both a magnitude (number and units) and a direction. When our displacement was equal to zero, there was no need to specify a direction because there was no overall change in position. However, when there is a vector quantity that is not equal to zero, a direction must be stated to describe the quantity completely.

In physics, vector quantities are often represented by arrows. The length of the arrow can indicate the relative magnitude or size of the vector. The direction of the arrow indicates the direction of the vector. Using arrows to represent vectors allows us to solve some problems graphically. These arrows also help us visualize what is going on in a particular problem.

To illustrate the use of arrows to represent vectors, and to reinforce the distinction between scalar and vector quantities, let's try a problem together.

Example 1

A squirrel starts from a reference point (origin) and travels 3.0 meters north to fetch an acorn. The squirrel then travels 5.0 meters east, and then another 3.0 m south. Find both the total distance the squirrel traveled, and its overall displacement.

Determining the distance that the squirrel traveled is quite easy, requiring neither arrows nor a class in physics to find. We simply add each of the distances together.

Total Distance = 3.0 m + 5.0 m + 3.0 m = 11.0 m

Notice that we can represent the total distance with numbers (11.0) and units (m) without mentioning direction, because *distance is a scalar quantity*. To illustrate the *displacement* of the squirrel we will represent each of the individual changes in position with **vector arrows**. The length of the arrows will relate to the distances the squirrel covered and the directions of the arrows will indicate the directions of the individual displacements. According to the standard convention, we will treat the top of the page as north, the bottom as south, the right as east, and the left as west.

Because each of these arrows represents a portion of the squirrel's overall displacement, they are called **component vectors**. When we add them all together to find the overall displacement, we will find the **resultant vector**. We graphically add the arrows together using what is called the "tip-to-tail" method. As the name of this method implies, component vectors are added together in such a way that the *tip* of one component is made to touch the *tail* of the next component. The order of adding the component vectors does not matter.

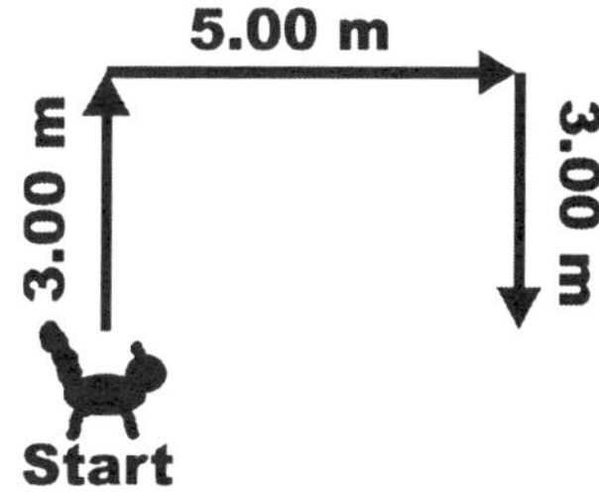

Figure 1.1

The *component vectors* are added "tip-to-tail."

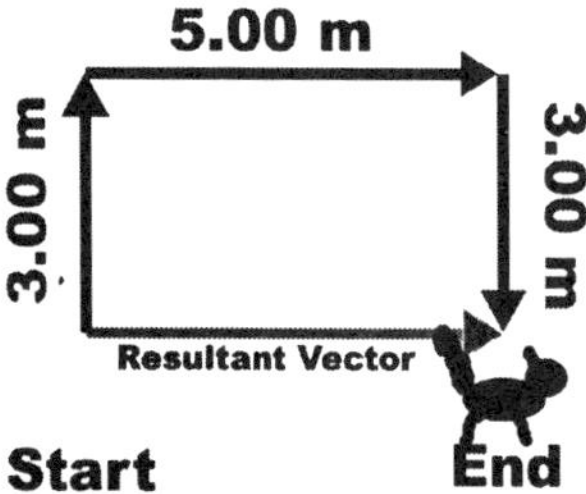

Figure 1.2

The next step in this method involves drawing the **resultant vector**. The resultant vector is the sum of all of the component vectors. Unlike the component vectors, the resultant vector is drawn from the *tail* of the first component vector to the *tip* of the last, as shown in Figure 1.2.

The *resultant vector* is drawn from "tail-to-tip."

As with the component vectors, the magnitude of the resultant vector is indicated by the length of the arrow, and the direction is also shown. The advantage of this graphical method of vector addition is that if you draw the component vectors to scale, you can measure the magnitude of the resultant vector with a ruler. Even without a ruler, you can see that the resultant vector is the same length and in the same direction as the component vector for 5.00 m to the east. So then 5.00 m is our resultant vector, or overall displacement.

Answer

Total Distance = 3.00 m + 5.00 m + 3.00 m = **11.0 m**

Total Displacement = **5.00 m east**

You may have noticed that the graphical method wasn't entirely necessary for solving the problem in Example 1. Many of these problems can be solved with simple algebraic addition. The key to solving these problems involves assigning algebraic signs to the different directions. For example, we can consider the direction forward as a positive displacement, and reverse as a negative displacement. For Example 1, let's consider north positive and south negative. Let's also designate east as positive and west as negative. Using these conventions, we can find the resultant displacement for the problem in Example 1 as follows:

Component Displacement	Algebraic Representation
d_1 = 3.0 m north	+3.0 m
d_2 = 5.0 m east	+5.0 m
d_3 = 3.0 m south	–3.0 m

Now, we only add the vectors that lie along the same axis, so we add vectors d_1 and d_3, and find that they add up to zero, as shown here:

$$d_1 + d_3 = +3.0\text{ m} + (-3.0\text{ m}) = 0\text{ m}$$

Because these two component vectors cancel each other out, the resultant vector is equal to the remaining component vector, **5.0 m to the east**.

Which method you use to solve a particular problem will often depend on the specifics of the problem, and how much time you have to solve it. One thing that you will notice in physics is that the problems often seem easier to solve if you draw a sketch of what is going on in the problem. Often, the most successful students are the ones that get in the habit of making a visual representation of the problem.

Let's try a couple more example problems. First, let's try a problem that you may want to solve algebraically. If the answer comes to you right away, you are ready to move on to harder problems.

Example 2

A football quarterback backpedals for 1.4 m before seeing an opening, and then runs forward for 3.5 m. Find his overall displacement.

Given: $d_1 = -1.4\text{ m}$ $d_2 = 3.5\text{ m}$

Find: d_{net}

Solution: $d_{net} = d_1 + d_2 = -1.4\text{m} + (+3.5\text{ m}) =$ **+2.1 m forward**

Example 2 represents the simplest of these types of problems, because all of the displacement occurred along a single axis, or a straight line. The only "trick" to this type of problem is to be sure to designate one direction as positive and the other as negative.

For our third example, let's try something a bit more difficult. Read the question first and then try to solve it on your own. I would recommend making a diagram of the action. Some people get this type of example right away, while others need to have it explained to them at first. The difference might have to do with how long it has been since you have studied geometry and trigonometry. If you have trouble with this next example, be sure to read the review of skills in the introduction of this book.

Example 3

A dog walks 3.0 m east and sniffs a fire hydrant. It then turns and walks 4.0 m north to sniff a tree. What is the overall displacement of the dog?

Notice that the component vectors do not lie along the same axis, so you can't just add them algebraically. The solution to this will only present itself if you visualize the actual component displacements in your mind or on paper. Try the problem before looking at my following explanation.

The first thing that we want to do is sketch the problem using the "tip-to-tail" method that I described earlier in this lesson. We draw the component vectors in such a way that the *tip* of one vector touches the *tail* of the other. It doesn't matter which vector you draw first, as I will show you shortly. Remember, if you use a ruler and draw these component vectors to scale, you will be able to find the magnitude of the resultant vector with the same ruler. For example, your scale may set each meter equal to a centimeter. You would then draw the first displacement vector (3.0 m east) 3.0 cm long, and the second displacement vector (4.0 m north) 4.0 cm long.

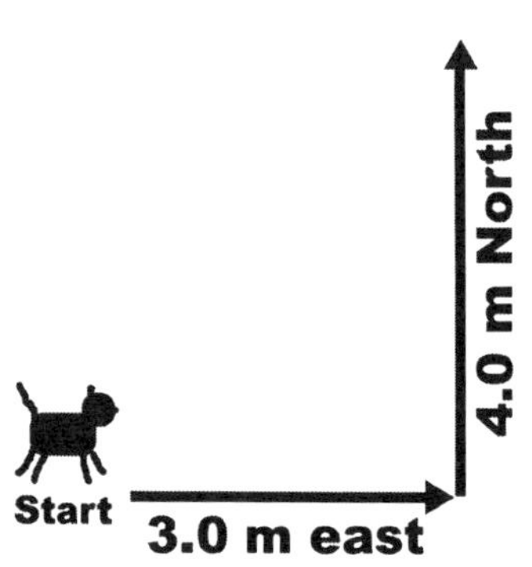

Figure 1.3

Now we are ready to draw the resultant vector. Remember that, unlike the component vectors, the resultant vector is drawn from the *tail* of the first component vector to the *tip* of the last, or "tail-to-tip."

Figure 1.3 is not drawn to scale. If it were, we could find the magnitude of the resultant vector by measuring the resultant vector and apply it to our original scale. If you were careful with your own diagram, you would find it to be 5.0 cm long, representing a resultant vector of 5.0 m. However, you would still need to determine the direction of the resultant vector.

If you didn't draw your diagram to scale, or you are not confident in your ability to find the magnitude of the resultant vector with the graphical method, you could also use the Pythagorean theorem.

The Pythagorean Theorem:

$$a^2 + b^2 = c^2$$

Given: $d_1 = 3.0$ m $\qquad d_2 = 4.0$ m

Find: d_r (resultant displacement)

Isolate:

Let's rewrite the Pythagorean theorem using our symbols:

$$d_r^{\,2} = d_1^{\,2} + d_2^{\,2}$$

Now, take the square root of both sides of the equation:

$$\sqrt{d_r^{\,2}} = \sqrt{d_1^{\,2} + d_2^{\,2}}$$

We get: $d_r = \sqrt{d_1^{\,2} + d_2^{\,2}}$

Solution:

$$d_r = \sqrt{d_1^{\,2} + d_2^{\,2}} = \sqrt{(3.0\,\text{m})^2 + (4.0\,\text{m})^2}$$
$$= \sqrt{9.0\,\text{m}^2 + 16\,\text{m}^2} = \sqrt{25\,\text{m}^2} = \mathbf{5.0\,m}$$

Whichever method we use to find the magnitude of the resultant vector, we will still need to find the direction. In some situations, it may be enough to indicate that the direction of the resultant vector is "north-east," but in other situations you will need to be more specific. If you carefully measured the lengths of the vectors involved, you can find the angle of the displacement vector using a protractor. If you are familiar with the type of triangle formed by our vectors, you may know the angles between each of the sides. If not, you can make use of one of the trigonometry functions that we went over in the Introduction.

$$\theta = \tan^{-1}\left(\frac{\text{opposite side}}{\text{adjacent side}}\right) = \tan^{-1}\left(\frac{4.0\,\cancel{\text{m}}}{3.0\,\cancel{\text{m}}}\right) = \mathbf{53.1^\circ}$$

We find that the overall displacement of the dog is **5.0 m at 53.1° north of east.**

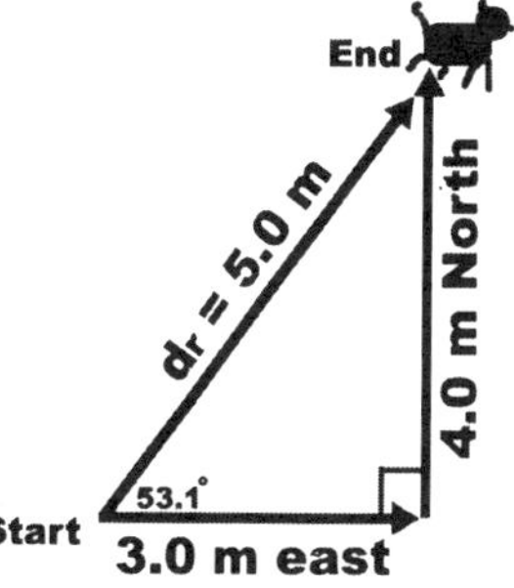

Figure 1.4

I wanted to mention one last thing before we move on to the practice questions for this section. Remember that we said it doesn't matter in what order you add component vectors, provided that you add them from tip-to-tail? Try the problem again, this time adding the tip of the 4.0 m vector to the tail of the 3.0 m vector. You will find the same resultant vector. In other words, it wouldn't matter if the dog went 4.0 m north and then 3.0 m east, or if it went 3.0 m east and then 4.0 m north. It still ends up in the same place.

Try the following practice problems. Check your answers in the Answer Key (page 73) at the end of the chapter before moving on to the next lesson.

Lesson 1–1 Review

1. _______________ is a vector quantity that describes the change in an object's position in terms of direction and distance.
2. A girl walks 33 m north, then 42 m east, and then 23 m west. Find the distance that she traveled.
3. A boy rides his bike 15 m due east and then 33 m due north. Find his displacement.

Lesson 1–2: Speed and Velocity

Just as displacement is the vector counterpart to the scalar quantity called distance, **velocity** is the vector counterpart to the scalar quantity called speed. While the speed of an object can be described with only numbers and units, the velocity of an object includes direction. In other words, velocity is the speed of an object in a particular direction. For this reason, it is possible for the speed of an object to remain constant, but still have its velocity change if it changes direction.

When motion is limited to a single axis, it is more convenient to indicate the direction the object is traveling with a + or – sign. For example, instead of saying that a car has a velocity of 25.0 m/s to the south, we can say that the car has a velocity of –25.0 m/s. It is important for you to get used to the concept of using a negative sign to indicate the direction. You can't have a negative speed, because nothing is going to move slower than stopped. A negative sign for velocity doesn't mean "slow" (for example, –25.0 m/s indicates a faster speed than, say, +12.0 m/s) it simply means velocity in the direction opposite of what is considered positive.

What you will learn in this lesson is that the techniques that we went over for determining the resultant displacement in Lesson 1–1, work for velocity vectors as well. In fact, they work for all of the vector quantities that you will encounter while you study physics.

Before we go over some example problems, let's make sure that you understand what velocity really is. What is the difference between displacement and velocity? Displacement is the change in an object's position in a particular direction. Velocity is that rate at which the displacement occurs. In other words,

$$\text{Velocity} = \frac{\text{displacement}}{\text{change in time}} = \frac{\text{change in position}}{\text{change in time}}.$$

Velocity is usually represented by the symbol v and is typically measured in meters/second (m/s), although any unit of length or time could be substituted. The formula for determining velocity is

$$v = \frac{d}{\Delta t} = \frac{x_f - x_i}{\Delta t}.$$

Instantaneous velocity is the velocity that an object has at a particular instant. **Average velocity** is the average velocity an object travels with over a period of time. These concepts should be easy to relate to their "speed" counterparts from everyday life. For example, the speedometer of a car shows the driver the *instantaneous speed*. If a driver covers 100 miles in 2 hours, his or her *average speed* was 50 miles per hour. When we say an object has **constant velocity**, we mean that its velocity is not changing. Remember, this means that the object is not speeding up, slowing down, *or* changing direction.

Let's go over a few examples that will demonstrate the variety of velocity problems that you might encounter. For our first example, let's try one that represents the simplest type of these problems.

Example 1

Find the average velocity of a ball if it rolls 30.0 m to the right in 5.0 seconds.

Given: d = 30.0 m $\Delta t = 5.0$ s

Find: v_{avg}

Solution: $v_{avg} = \frac{d}{\Delta t} = \frac{30.0\text{ m}}{5.0\text{ s}} =$ **6.0 m/s to the right**

Example 1 is simply a matter of identifying which variables you have been given and plugging them into the proper formula. Notice that our answer includes numbers, units, and a direction. Now we will try a problem that is marginally harder. For this problem you will need to be able to isolate the unknown.

Example 2

A car moves due north with a constant velocity of 30.0 m/s. How long will it take this car to travel 110 m to the north?

Given: v = 30.0 m/s   d = 110 m

Find: Δt

Notice that the car is moving at constant velocity, and, in such cases, the instantaneous velocity is the same as the average velocity. We can use the same formula that we used in our last example, but we must isolate Δt.

Isolate:

Original formula: $v = \frac{d}{\Delta t}$

Multiply both sides by Δt: $\Delta t \times v = \frac{d}{\cancel{\Delta t}} \times \cancel{\Delta t}$

Divide both sides by v: $\frac{\Delta t \times \cancel{v}}{\cancel{v}} = \frac{d}{v}$

We get: $\Delta t = \frac{d}{v}$

Solution: $\Delta t = \frac{d}{v} = \frac{110\ \cancel{\text{m}}}{30.0\ \cancel{\text{m}}/\text{s}} =$ **3.7 s**

For our next example, we will cover a classic example of a problem that is designed to test whether or not a student differentiates between scalar and vector quantities. Read the question carefully and make sure that you understand what is being asked.

Example 3

A race car travels completely around a circular track, covering a distance of 850 m in 25 s before stopping at the spot at which it started. Determine the average velocity of the car during this period of time.

Given: d = 0 m $\Delta t = 25$ s

Find: v_{avg}

Solution: $v_{avg} = \frac{d}{\Delta t} = \frac{0\text{ m}}{25.0\text{ s}} = \mathbf{0\ m/s}$

Example 3 is what students call a "trick" question, and indeed, it isn't entirely fair. Did the answer to Example 3 surprise you? Perhaps you thought that the answer was 34 m/s? The trick to this question has to do with the difference between speed and velocity. Velocity is based on displacement, and the final displacement of the race car is zero, because it ended up in the same spot in which it started. The average speed of the car, which is based on the distance traveled rather than the displacement, is equal to 34 m/s, but the average velocity is zero.

For our final example, let's try a problem that is sometimes called a relative motion problem. Before I introduce the problem, let me explain what I mean by relative motion. Suppose I told you that it is so easy to throw a 90-mile an hour fastball that my 4-year-old daughter could do it. Would you believe it? What if she was sitting in the back of a van that was traveling 75 miles per hour north on a highway? If she threw a ball with a speed of 15 miles per hour north relative to the other people in the van in the same direction that the van was traveling, it would travel (75 miles/hour + 15 miles/hour) 90 miles per hour north relative to an observer standing on the road. Because the ball seems to travel at a different velocity, depending on whether or not you are inside the van, we say that its motion is relative to where the observer is.

Now, for our example.

Example 4

A child is riding in a car with a velocity of 25.0 m/s due east. The child throws a ball out the window with a velocity of 6.00 m/s south, relative to himself. What is the instantaneous velocity of the ball relative to the road?

Given: v_{car} = 25.0 m/s east v_{toss} = 6.00 m/s south

Find: v_r

Some people may be tempted to simply add the velocity of the car and the velocity at which the child threw the ball, but these vectors lie along different axes and are actually perpendicular to each other. What we need to do is add the component vectors in the same way that we added perpendicular component vectors in Lesson 1–1. When in doubt, sketch the problem out. You will see that we can solve the problem using the "tip-to-tail" method.

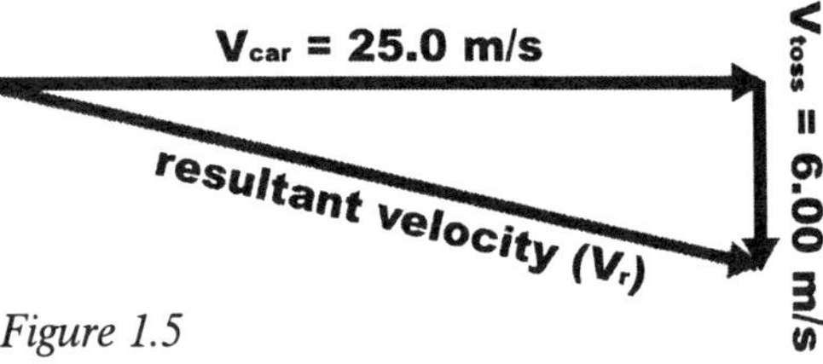

Figure 1.5

We can find the resultant velocity of the ball using the Pythagorean theorem.

$$c = \sqrt{a^2 + b^2} = \sqrt{(25.0\text{ m/s})^2 + (6.00\text{ m/s})^2} = \mathbf{25.7\text{ m/s}}$$

We can find the angle of this resultant velocity of the ball as shown here:

$$\theta = \tan^{-1}\left(\frac{\text{opposite side}}{\text{adjacent side}}\right) = \tan^{-1}\left(\frac{6.00\text{ m/s}}{25.0\text{ m/s}}\right) = \mathbf{13.5°}$$

Putting this information together, we have an initial velocity of **25.7 m/s at 13.5° south of east.**

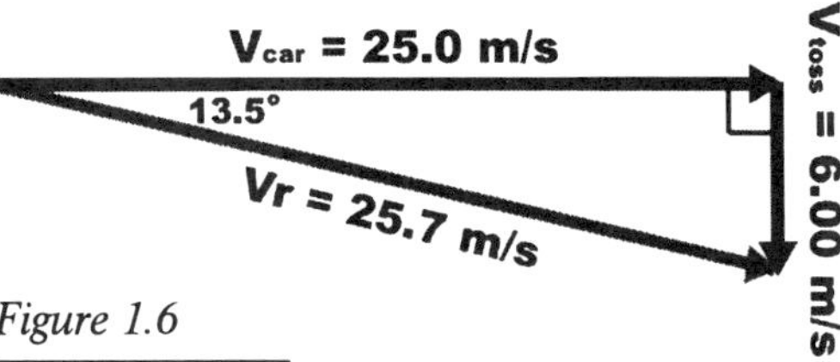

Figure 1.6

Try the following practice problems and check your answers at the end of the chapter before moving on to the next lesson.

Lesson 1–2 Review

1. __________ is the speed of an object in a particular direction.
2. Is it possible for a car to move with a constant speed, and still change its velocity?
3. An airplane with a velocity of 125 m/s south encounters a crosswind of 35 m/s to the west. What is the resultant velocity of the plane, relative to the ground?

Lesson 1–3: Acceleration

Just as the term velocity refers to the rate at which position changes, **acceleration** is the rate at which velocity changes. In other words,

$$\text{acceleration} = \frac{\text{change in velocity}}{\text{change in time}}.$$

Acceleration, which we will represent with the symbol a, is measured in m/s^2. The formula for velocity is

$$a = \frac{\Delta v}{\Delta t} = \frac{v_f - v_i}{\Delta t}.$$

When an advertisement claims that a car can go from "zero to 60 in six seconds," it is talking about acceleration. The term acceleration means approximately the same thing in physics as it does in everyday life with the exception that it is a vector quantity, so direction is important. If a car maintains a speed of 20.0 miles per hour as it goes around a turn, the speed is not changing, but the velocity is. Therefore, even a car moving at constant speed is said to be accelerating as it goes around a turn.

Like velocity, we can talk about the **instantaneous acceleration**, the acceleration at a particular instant, or the **average acceleration** that an object experiences over a period of time. If an object accelerates at a constant rate, we say that it moves with **constant**, or **uniform**, **acceleration.** When an object's acceleration is constant, its instantaneous acceleration is the same as its average acceleration for that period of time.

One potential confusing aspect of working with acceleration is the fact that there is a difference between a **deceleration** (slowing down) and a **negative acceleration** (acceleration in the direction that has been designated as negative). Imagine a car that is heading north with a velocity of 25 m/s. We may consider north to be positive, so this would represent a positive velocity. If this car then turned around and headed south, we would consider the new velocity to be negative. If the car speeds up to the south, it would experience a negative acceleration, and the car would be going faster and faster!

When an object accelerates in the same direction as its velocity, it speeds up. When the acceleration is in the opposite direction as its velocity, the object will slow down.

Let's take a look at the variety of acceleration questions that can be solved using the basic formula.

Example 1

A car driving in a straight line accelerates from 12.0 m/s to 23.0 m/s in a period of 3.0 s. Find the average acceleration experienced by the car during this period of time.

Given: $v_i = 12.0$ m/s $\quad v_f = 23.0$ m/s $\quad \Delta t = 3.0$ s

Find: a_{avg}

Solution: $a_{avg} = \dfrac{v_f - v_i}{\Delta t} = \dfrac{23.0\text{ m/s} - 12.0\text{ m/s}}{3.0\text{ s}} = \dfrac{11.0\text{ m/s}}{3.0\text{ s}} = \mathbf{3.7\ m/s^2}$

Notice that in Example 1 we had a car with a positive initial velocity and a positive acceleration, and the speed of the car increased. Let's look at an example with a negative acceleration.

Example 2

A car with an initial velocity of 35.0 m/s experiences a constant acceleration of –7.0 m/s². How long does it take the car to come to a complete stop?

Given: $v_i = 35.0$ m/s $\quad v_f = 0.0$ m/s $\quad a = -7.0$ m/s²

Find: Δt

Solution: $\Delta t = \dfrac{v_f - v_i}{a} = \dfrac{0.0\text{ m/s} - 35.0\text{ m/s}}{-7.0\text{ m/s}^2} = \dfrac{-35.0\ \cancel{\text{m}}/\text{s}}{-7.0\ \cancel{\text{m}}/\text{s}^2} = \mathbf{5.0\ s}$

Example 2 shows how important it is to subtract the initial velocity from the final velocity ($\Delta v = v_{final} - v_{initial}$) to find the change in velocity. Some students would reverse this calculation, subtracting the final velocity from the initial. In a problem such as Example 2, this would yield a positive value for Δv and a negative value for Δt! Unless your teacher gives you a problem on time travel, and I wouldn't put that past some physics teachers, you probably shouldn't end up with a negative change in time.

Now, let's try an example with a negative velocity and a positive acceleration, just to reinforce the importance of the algebraic signs. Try this problem on paper by yourself before looking ahead to my answer.

Example 3

A motorcycle with an initial velocity of –21.0 m/s experiences a constant acceleration of 4.00 m/s^2 for a period of 3.0 s, what will be the final velocity of the motorcycle?

Given: $v_i = -21.0$ m/s a = 4.00 m/s^2 $\Delta t = 3.0$ s

Find: v_f

The first key to Example 3 is to note that the initial velocity is negative, but the acceleration is positive. This will result in the motorcycle slowing down, not speeding up. If you end up with a greater final velocity, then you probably mixed up the algebraic signs. The other thing that some people might find challenging is the isolation of the unknown, v_f, as shown here:

Isolate:

Original formula: $a = \dfrac{v_f - v_i}{\Delta t}$

Multiply both sides by Δt and it cancels on the right side:

$$\Delta t \times a = \frac{v_f - v_i}{\cancel{\Delta t}} \times \cancel{\Delta t}$$

This gives us: $\Delta t \times a = v_f - v_i$

Now, add v_i to both sides of the equation, and cancel:

$$\Delta t \times a + v_i = v_f - \cancel{v_i} + \cancel{v_i}$$

Rearranging the formula, we get: $v_f = v_i + a\Delta t$

This working formula, $v_f = v_i + a\Delta t$, is one that you should become familiar with. It will be useful for solving a large number of problems. Now that we have this working formula, we are ready to solve example 3. Compare your answer to the one shown here:

Solution:
$$v_f = v_i + a\Delta t = -21.0\text{ m/s} + (4.00\text{ m/s}^2)(3.0\text{ s}) = -21.0\text{ m/s} + 12\text{ m/s} = \mathbf{-9.0\text{ m/s}}$$

Notice that the motorcycle experiences a positive acceleration, and slows down. This occurs because the velocity and acceleration were in opposite directions. Let's try one more example together, before moving on to the practice problems.

Example 4

A horse starts from rest and accelerates at an average rate of –2.5 m/s² over a period of 7.0 s. What is the final velocity of the horse after this time?

Given: $v_i = 0.0$ m/s a = –2.5 m/s² Δt = 7.0 s

Find: v_f

Solution: $v_f = v_i + a\Delta t = 0.0\text{ m/s} + (-2.5\text{ m/s}^2)(7.0\text{ s}) = \mathbf{-18\text{ m/s}}$

Does the negative value for the horse's velocity in Example 4 imply that the horse is running backwards? No, it is simply heading in the direction opposite to what is considered positive in this particular problem.

Now let's move on to the practice problems. Remember to check your answers with the Answer Key.

Lesson 1–3 Review

1. ______________ is the rate at which an object's velocity changes.
2. An airplane with an initial velocity of 0.0 m/s experiences a uniform acceleration of 9.5 m/s². How much time will it take to reach a velocity of 125 m/s?
3. Is it possible for an object to speed up while experiencing a negative acceleration?

Lesson 1-4: Constant Acceleration

In situations where the acceleration of an object is constant (including when the acceleration is zero) we can use the following group of equations to find a missing variable, such as displacement or change in velocity. These equations are so important and useful that you should become very familiar with them. Even if your instructor doesn't require you to memorize them, you will want to make sure that you can isolate any of the potential unknowns in each of them. Many problems begin with the object at rest, in which case you can use the simplified version of the equations on the right.

Equations for Motion With Constant Acceleration

When the Initial Velocity Is Not Zero	When the Initial Velocity Is Zero
$v_f = v_i + a\Delta t$	$v_f = a\Delta t$
$v_f^2 = v_i^2 + 2ad$	$v_f^2 = 2ad$
$d = v_i\Delta t + \frac{1}{2}a\Delta t^2$	$d = \frac{1}{2}a\Delta t^2$
$d = \left(\frac{v_i + v_f}{2}\right)\Delta t$	$d = \left(\frac{v_f}{2}\right)\Delta t$

The ability to select a proper formula for a given situation is critical. Beginners will often wonder, *Which formula do I use?* The answer is really quite simple. First, you look at which quantities you are given in the situation and which quantity is your unknown. Then, you select a formula for which you have all of the variables except for the quantity that you want to solve for. You will encounter many situations where a problem can be approached from different angles and solved with different formulas, but the key will always be in identifying a formula that will bring you closer to where you want to get.

Let's go over our first example.

Example 1

A car with an initial velocity of 8.00 m/s accelerates at a constant rate of 3.00 m/s^2 until it reaches a velocity of 24.0 m/s. How far does the car travel in this time?

Before you start searching for the proper formula for solving the problem, slow down and take the time to make note of which quantities you have been given and what the unknown is. Some students feel like this step is a waste of time at first, but they come to realize that it actually saves them time by giving them a systematic approach to solving any problem. You must be careful to interpret the word problems correctly in order to accurately identify the given quantities. In these situations, knowing which units are associated with each quantity (for example, acceleration is measured in m/s^2) can be a big help.

Once you have a list of the quantities involved in the problem, selecting the proper formula should be easy, as shown here:

Given: $v_i = 8.00$ m/s $v_f = 24.0$ m/s a = 3.00 m/s^2

Find: d

Based on what we have been given and what we want to find, the formula we select should be $v_f^2 = v_i^2 + 2ad$. For some, isolating the unknown is a challenge, so I will show the isolation for displacement.

Isolation:

Original formula: $v_f^2 = v_i^2 + 2ad$

Subtract v_i^2 from both sides: $v_f^2 - v_i^2 = \cancel{v_i^2} + 2ad - \cancel{v_i^2}$

This gives us: $v_f^2 - v_i^2 = 2ad$

Now, divide both sides by 2a: $\frac{v_f^2 - v_i^2}{2a} = \frac{\cancel{2a}d}{\cancel{2a}}$

This gives us our working formula: $d = \frac{v_f^2 - v_i^2}{2a}$

We are now ready to solve the problem.

Solution:

$$d = \frac{v_f^2 - v_i^2}{2a} = \frac{(24.0\text{ m/s})^2 - (8.00\text{ m/s})^2}{2(3.00\text{ m/s}^2)} = \frac{512\text{ m}^2/\cancel{\text{s}}^2}{6.00\text{ }\cancel{\text{m}}/\cancel{\text{s}}^2} = \mathbf{85.3\text{ m}}$$

Now, let's show a couple of more examples. Try to solve each of the following problems on your own, before checking my solutions.

Example 2

A rocket sled starts from rest and experiences a uniform acceleration of 22.5 m/s² for 4.00 s. What will be the final velocity of the sled after this period of time?

Given: $v_i = 0.0$ m/s a = 22.5 m/s² Δt = 4.00 s

Find: v_f

Solution: $v_f = a\Delta t = (22.5 \text{ m/s}^2)(4.00 \text{ s}) = \mathbf{90.0\ m/s}$

Example 2 is a problem with what I call an "implied" given. When a problem states that an object "starts from rest" what it really tells us is that the initial velocity is equal to zero. Look for this common phrase when you search a word problem for the given information.

Also, notice that we used one of the formulas for when the initial velocity is zero. If you started with the version of the formula that included the initial velocity, you would have ended up with the same answer, because the value for it was zero as shown below.

$$v_f = v_i + a\Delta t = 0.0 \text{ m/s} + (22.5 \text{ m/s}^2)(4.00 \text{ s}) = \mathbf{90.0\ m/s}$$

Starting with the version of the formula where the initial velocity has been removed simply saves you a step.

Example 3

A bus has an initial velocity of 19.0 m/s. The driver slams on the brakes and the bus experiences a uniform acceleration until it comes to a stop. If the stopping distance (d) of the bus is 25.0 m, find its acceleration.

Given: $v_i = 19.0$ m/s $v_f = 0.00$ m/s d = 25.0 m

Find: a

Isolation:

Original formula: $v_f^2 = v_i^2 + 2ad$

Subtract v_i^2 from both sides: $v_f^2 - v_i^2 = \cancel{v_i^2} + 2ad - \cancel{v_i^2}$

Because the final velocity is zero, v_f^2 is also zero.

We can cross it out: $\cancel{v_f^2} - v_i^2 = 2ad$

Now, we divide both sides by 2d: $\frac{-v_i^2}{2d} = \frac{\cancel{2}a\cancel{d}}{\cancel{2d}}$

This gives us our working formula: $a = \frac{-v_i^2}{2d}$

Solution: $a = \frac{-v_i^2}{2d} = \frac{-(19.0\text{ m/s})^2}{2(25.0\text{ m})} = \frac{-361\text{ m}^2/\text{s}^2}{50\text{ }\cancel{\text{m}}} = \mathbf{-7.22\text{ m/s}^2}$

The negative sign in our answer shows that the acceleration is in the direction opposite to the initial velocity, which makes sense because the bus was slowing down. Let's try one more example together, before you move onto the practice problems.

Example 4

A go-cart with an initial velocity of 1.0 m/s accelerates uniformly at a rate of 0.25 m/s² for a period of 12.0 s. What distance will the go-cart travel during this time?

Given: $v_i = 1.0$ m/s $a = 0.25$ m/s² $\Delta t = 12.0$ s

Find: d

Solution:

$$d = v_i\Delta t + \frac{1}{2}a\Delta t^2 = (1.0\text{ m/s})(12.0\text{ s}) + \frac{1}{2}(0.25\text{ m/s}^2)(12.0\text{ s})^2$$
$$= \mathbf{30.\text{ m}}$$

Try these review problems.

Lesson 1–4 Review

1. A bus starts at rest and accelerates at a rate of 4.5 m/s². What will be the velocity of the bus after 3.8 s?
2. An airplane starts at rest and experiences uniform acceleration until it reaches 45.0 m/s. If it travels 55.0 m during this time, find the acceleration.
3. A car with an initial velocity of 5.00 m/s accelerates uniformly to 25.0 m/s in a period of 5.00 s. How far does the car travel during this period of time?

Lesson 1–5: Free Fall

Think about some of the experiences that you have had with falling objects. Have you ever dropped an object, such as a glass, and been relieved when it didn't break? Did you ever drop water balloons out of a high window? Did you ever make a connection between how high a baseball flies, and how much it stings your hand when you catch it?

Why is it that when you throw a baseball a few feet into the air you can catch it with your bare hand and not feel a sting, but when a baseball falls from a much higher point, like when someone hits a pop fly, the ball hits your hand hard enough to hurt? The answer, of course, is that the ball is falling faster when it has fallen a greater distance. This is clear evidence that objects accelerate (speed up) as they fall.

Can you imagine what the world would be like if objects didn't accelerate as they fell? If objects maintained a constant velocity as they fell, then jumping off a 50-story building would be the same as jumping down from the curb! Dropping a glass from an inch above the floor would be the same as dropping one off a cliff! But, of course, falling objects *do* accelerate.

Why do objects accelerate as they fall? You will learn more about this in our next chapter, as we study Newton's second law of motion. The force of gravity causes objects to accelerate, but we said that we wouldn't discuss forces in this chapter. We won't discuss *why* the objects accelerate yet, but we will discuss *how* they accelerate.

When objects are made to fall close to the surface of Earth, where the only unbalanced force acting on them is gravity, we say that they are "falling freely" or experiencing **free fall**. Gravity will cause such objects to accelerate at a constant rate of 9.81 m/s^2. This rate of acceleration is a constant (near the surface of Earth) called "the acceleration due to gravity," and is given the symbol g. Be careful not to confuse the symbol for the acceleration due to gravity with the symbol for grams.

Notice that there is a difference between constant velocity and constant acceleration. If an object has a constant velocity, it means that it doesn't speed up, slow down, or change direction. An object that has a constant nonzero acceleration does speed up, slow down, and/or change direction. When we say that an object is accelerating at a rate of 9.81 m/s^2, we mean that every second that it is falling, its velocity changes by 9.81 m/s. If you throw a ball upward, it will experience this acceleration during its entire flight. On the way up, its velocity is decreasing at a rate of 9.81 m/s every second, until it reaches a velocity of zero. After that, it will begin to come back down, and its velocity will be increasing by 9.81 m/s every second. It is very important to remember that its acceleration doesn't change during the entire flight (g is a constant) but its velocity certainly does.

We have spent some time in this chapter discussing the idea of using signs for direction. It is a fairly standard practice for physics students and teachers to designate "upward" as positive and "downward" as negative. Using this sign convention, that means that we will use the value of $-9.81\ m/s^2$ for g.

The Acceleration Due to Gravity Close to the Surface of Earth

$$g = -9.81\ m/s^2$$

It is important for you to remember that g, the acceleration due to gravity, is just a specific acceleration. This means that you can use any of the formulas for uniform acceleration for free-fall problems. You just substitute g for a. Some students and teachers prefer to rewrite the formulas to include the g, as shown here, to remind them to use the value of $-9.81\ m/s^2$ for the acceleration:

Equations for Free-Fall, With Constant Acceleration of g

When the Initial Velocity Is Not Zero	When the Initial Velocity Is Zero
$v_f = v_i + g\Delta t$	$v_f = g\Delta t$
$v_f^2 = v_i^2 + 2gd$	$v_f^2 = 2gd$
$d = v_i\Delta t + \frac{1}{2}g\Delta t^2$	$d = \frac{1}{2}g\Delta t^2$
$d = \left(\frac{v_i + v_f}{2}\right)\Delta t$	$d = \left(\frac{v_f}{2}\right)\Delta t$

Of course, there is really no need to memorize or even rewrite a new set of equations. Substituting g for a is simply done to remind us that the acceleration for each of these problems is a constant, which we can count among our "given" quantities. You will find that the problems that we cover in this lesson are really just special examples of the uniform equations that we covered in Lesson 1–4. The only difference is that the acceleration is not usually specifically stated in these problems, because it is a constant.

Let's try some examples together.

Example 1

A boy drops a rock into well and times how long it takes for him to hear it strike the water below. If the rock takes 2.1 seconds to hit the water, how deep is the well?

Ah, the classic "boy drops a rock in a well problem." Despite the fact that many of today's physics students may never see a well, this classic never seems to go out of style. One key to solving these problems is proper interpretation. When the problem states that a boy "drops" a rock it implies that the rock is starting from rest, with an initial velocity of zero. If the boy "threw" the rock, it would be a different story, and an initial velocity would be included in the problem. The other key is to remember that once the boy releases the rock, the only unbalanced force acting on it (when we ignore friction) is the force of gravity. Because this is an example of an object in free fall, the acceleration that it experiences will be equal to g, $-9.81\ m/s^2$. Given all of this, can you solve the problem before checking my answer?

Given: $v_i = 0.0\ m/s$ $\Delta t = 2.1\ s$ $g = -9.81\ m/s^2$

Find: d

Solution: $d = \frac{1}{2} g \Delta t^2 = (0.5)(-9.81\ m/s^2)(2.1\ s)^2 = -21.6\ m = \mathbf{-22\ m}$

It might surprise you to find that the displacement has a negative value, but it shouldn't be such a shock. Remember, we have decided to designate "downward" as the negative direction, so the negative sign in our answer tells us that the rock fell downward, rather than upward. Isn't it interesting that you can use this technique in real life to determine the height or depth of something? The physics that you learn has all kinds of practical applications.

Let's try an example with an object going upward. Read the question and try the problem on your own before checking my answer.

Example 2

An archer points her bow straight upward and fires an arrow with an initial velocity of 22.0 m/s. What is the maximum height that the arrow will reach?

For this type of problem, you need to remember that the arrow will have a positive (upward) initial velocity, but a negative (downward) acceleration. This means that the arrow will go slower and slower as it goes higher and higher. Eventually, the velocity of the arrow will become zero, at the instant that it reaches its maximum height (max d_y), or displacement. After that, the arrow starts to fall back down, accelerating along the way. So, that archer best get out of the way! The key to this type of problem is to use the value of zero for the final velocity, even though it is only the final velocity for the upward motion of the arrow. Sometimes I will call it v_f (at max d_y) to remind myself that the arrow still needs to fall back down.

Given: $v_i = 22.0$ m/s v_f (at max d_y) = 0.00 m/s $g = -9.81$ m/s^2

Find: d

Isolation:

We start with the formula: $v_f^2 = v_i^2 + 2gd$

Subtract v_i^2 from both sides of the equation:

$$v_f^2 - v_i^2 = \cancel{v_i^2} - \cancel{v_i^2} + 2gd$$

Which gives us: $v_f^2 - v_i^2 = 2gd$

Now we divide both sides of the equation by 2g: $\frac{v_f^2 - v_i^2}{2g} = \frac{\cancel{2g}d}{\cancel{2g}}$

Rearranging the equation, we get: $d = \frac{v_f^2 - v_i^2}{2g}$

Solution:

$$d = \frac{v_f^2 - v_i^2}{2g} = \frac{(0.0\text{ m/s})^2 - (22.0\text{ m/s})^2}{2(-9.81\text{ m/s}^2)}$$

$$= \frac{-484\text{ m}^2/\cancel{\text{s}^2}}{-19.62\ \cancel{\text{m}/\text{s}^2}} = \mathbf{24.7\ m}$$

Many problems that you will encounter in physics will have multiple parts. For example, we can use the problem from Example 2 and find the time that it takes the arrow to reach its maximum height. You might be tempted to use the value for d, which we just calculated, as one of the givens for the second part of the problem, but I am going to caution you not to for two reasons. First, the value that we have for d has been rounded already, and second, it is possible that we might have made an error in our calculations. Whenever possible, try to avoid using the answers from one part of a problem to solve the next part of a problem. If you can select a formula that allows you to solve the second part without relying on prior calculations, it is advisable to do so. You won't always have this option, but when you do, I suggest that you take it.

Example 3

An archer points her bow straight upward and fires an arrow with an initial velocity of 22.0 m/s. How many seconds will it take the arrow to reach its maximum height?

Given: $v_i = 22.0$ m/s v_f (at max d_y) = 0.00 m/s $g = -9.81$ m/s^2

Find: Δt

Isolate:

Starting with: $v_f = v_i + g\Delta t$

We subtract v_i from both sides: $v_f - v_i = \cancel{v_i} - \cancel{v_i} + g\Delta t$

Giving us: $v_f - v_i = g\Delta t$

Now, divide both sides by g: $\dfrac{v_f - v_i}{g} = \dfrac{g\Delta t}{\cancel{g}}$

Rearranging the equation, we get: $\Delta t = \dfrac{v_f - v_i}{g}$

Solution: $\Delta t = \dfrac{v_f - v_i}{g} = \dfrac{0.0 \text{ m/s} - 22.0 \text{ m/s}}{-9.81 \text{ m/s}^2} = \dfrac{-22.0 \cancel{\text{m}}/\cancel{\text{s}}}{-9.81 \cancel{\text{m}}/\text{s}^2} = \mathbf{2.24\ s}$

Keep in mind that the change in time that we just solved for only represents the amount of time that it takes for the arrow to reach its maximum height. Because the acceleration on the object is constant, it will require the same amount of time to fall back down to its original height (the height from which it was fired). If you were asked to determine the total time of flight, you would need to multiply the change of time that we calculated by 2, to take into account the time it takes for the arrow to fall back down. Of course, in a slightly harder type of problem, you might be asked to take the height of the archer into account.

For our last example together, let's calculate the final velocity of a falling object that has been thrown downward with an initial velocity in the negative direction. Although a force is initially exerted on such an object, after the object leaves the hand it is influenced only by gravity (when we ignore air resistance) so this, too, is an example of free fall.

Example 4

A child throws a rock with an initial velocity of –8.00 m/s from a window to strike the ground 30.0 m below. With what velocity will the rock strike the ground?

Given: $v_i = -8.00$ m/s d = –30.0 m g = –9.81 m/s²

Find: v_f

Isolate:

Starting with: $v_f^2 = v_i^2 + 2gd$

Simply take the square root of both sides: $\sqrt{v_f^2} = \sqrt{v_i^2 + 2gd}$

We get: $v_f = \sqrt{v_i^2 + 2gd}$

Solution:

$$v_f = \sqrt{v_i^2 + 2gd}$$
$$= \sqrt{(-8.00\text{ m/s})^2 + 2(-9.81\text{ m/s}^2)(-30.0\text{ m})}$$
$$= 25.54603687\text{ m/s}$$

Which rounds to **25.5 m/s**.

Lesson 1–5 Review

1. _______________ is the condition of an object that is falling with uniform acceleration caused by gravity.
2. When a baseball player hits a pop fly, what is the velocity and the acceleration of the ball when it reaches its highest point?
3. If a bullet were fired straight up with an initial velocity of 80.0 m/s, how high would it go? (Ignore air resistance.)

Lesson 1–6: Graphing Motion

Graphs are important tools for physics. If you can interpret graphs correctly, you will be able to extract a great deal of information from them quite quickly. Being able to construct graphs properly will give you the ability to summarize results in a format that is easy to read. This lesson will focus on the interpretation of motion graphs.

The first type of graph that we will study is called a **position vs. time graph.** Perhaps the quickest way to learn how to interpret this type of graph is to construct one. Let's go back to studying the motion of our friendly squirrel from Lesson 1–1. For the sake of simplicity, let's imagine that the squirrel is capable of changing its velocity instantaneously, so that it takes it no time to accelerate between two different velocities. For any period of time, the velocity of the squirrel will be uniform, which means that its instantaneous velocity will be the same as its average velocity for that period of time.

Let's imagine that when we first observe the squirrel it is at the base of a tree (the origin, or starting point). As we watch it, it moves in a straight line with uniform velocity for 2.0 seconds, at which point it stops at a distance of 5.0 m from the origin. We could graph the motion of the squirrel during this time as shown in Graph 1.1.

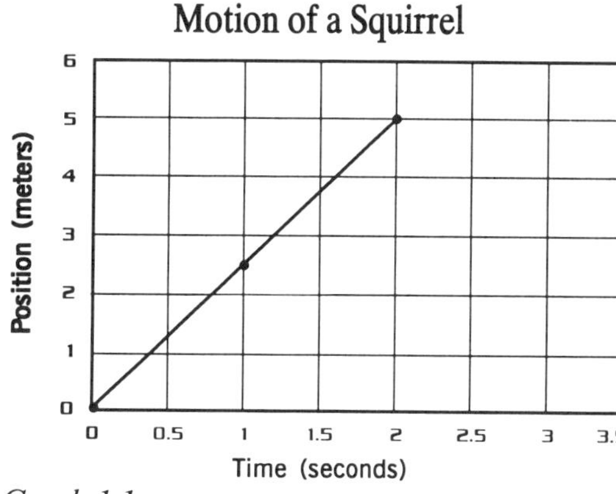

Graph 1.1

Notice that we can determine the position of the squirrel at any point in time by tracing back to the appropriate axis. For example, at 1.0 second, the squirrel was

2.5 meters away from the tree. We can also determine the average velocity of the squirrel, or the instantaneous velocity for any point if the velocity is uniform, during this period of time by finding the slope of the line! We simply pick two convenient points on the straight line segment, such as (0,0) and (2,5), and put them into the slope formulas, as shown here:

$$v_{avg} = \text{slope} = \frac{\Delta \text{y axis}}{\Delta \text{x axis}} = \frac{\text{displacement}}{\text{change in time}} = \frac{5\text{ m} - 0\text{ m}}{2\text{ s} - 0\text{ s}} = \frac{5\text{ m}}{2\text{ s}} = \mathbf{2.5\ m/s}$$

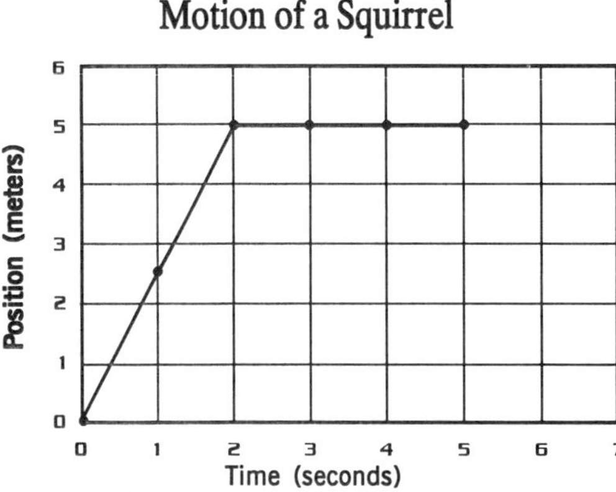

Graph 1.2

Let's imagine that the squirrel stops at this point and stays still for a period of 3.0 seconds. Let's add this data to our graph and see what it looks like (Graph 1.2).

We can see that the line representing the period of time when the squirrel was motionless (from 2.0 s to 5.0 s) is represented by a horizontal line. This makes sense because the slope of this line segment, and therefore the average velocity of the squirrel during this period of time, will be zero, as shown here:

$$v_{avg} = \text{slope} = \frac{\Delta \text{y axis}}{\Delta \text{x axis}} = \frac{\text{displacement}}{\text{change in time}} = \frac{5\text{ m} - 5\text{ m}}{5\text{ s} - 2\text{ s}} = \frac{0\text{ m}}{3\text{ s}} = \mathbf{0.0\ m/s}$$

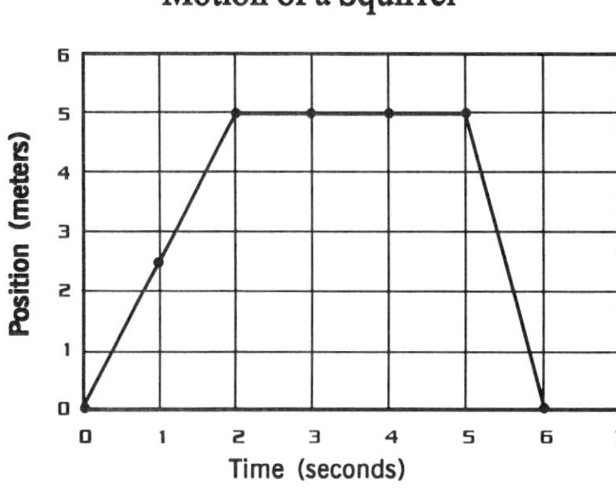

Graph 1.3

Next, let's imagine that our squirrel is startled by a loud noise, and sprints back to the tree in 1.0 s (this is a really fast squirrel!). Adding this part to our graph would give us the one shown in Graph 1.3.

As the squirrel returns to the tree (origin), the data creates a line segment with a negative slope and, therefore, a negative average velocity as the following equation shows.

$$v_{avg} = \text{slope} = \frac{\Delta\text{y axis}}{\Delta\text{x axis}} = \frac{\text{displacement}}{\text{change in time}} = \frac{0\text{ m} - 5\text{ m}}{6\text{ s} - 5\text{ s}} = \frac{-5\text{ m}}{1\text{ s}} = \mathbf{-5.0\ m/s}$$

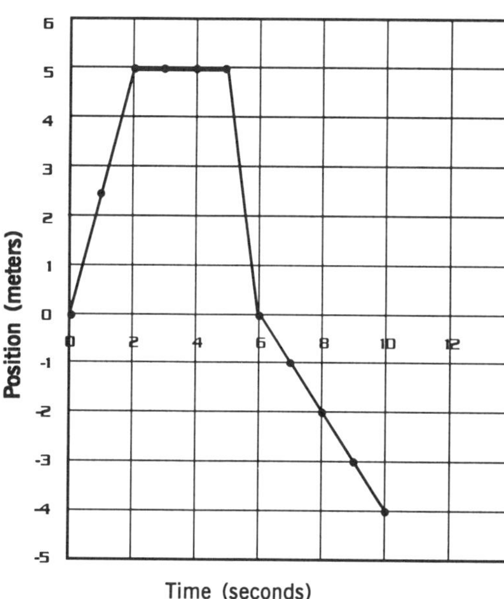

Graph 1.4

What does the negative velocity represent? A negative velocity is simply a velocity in the direction that is *opposite* to the direction that we called positive. The squirrel went back to the tree.

If the squirrel goes past the origin, beyond the tree, this line of his motion will dip below the *x*-axis, as shown in the next graph. The slope of the line segment would, of course, depend on how fast the squirrel went. For the sake of argument, let's imagine that the squirrel travels with a velocity of –1.0 m/s during this time (Graph 1.4).

Let's summarize what we have learned about position vs. time graphs.

Position vs. Time Graphs

1. Uniform velocity is represented by a straight line segment.
2. For uniform velocity, the instantaneous velocity is the same as the average velocity.
3. The slope of a line segment shows the average velocity of the object during a period of time.
4. A positive slope represents a positive velocity.
5. A horizontal line with a slope of zero represents a velocity of zero.
6. A negative slope represents a negative velocity.
7. When the line drops below the horizontal axis, the object moves past the origin, heading in the direction designated as negative.

The next type of graph that you will want to be able to create and read is a velocity vs. time graph. For fun, let's imagine that the students of an advanced physics course have just completed a robot that they have been working on, and they bring it to the gym to test its range of motion. They hit the On button, and the robot goes berserk! A student collects data about the velocity of the robot and wants to summarize the data on a graph, as shown in the following paragraphs.

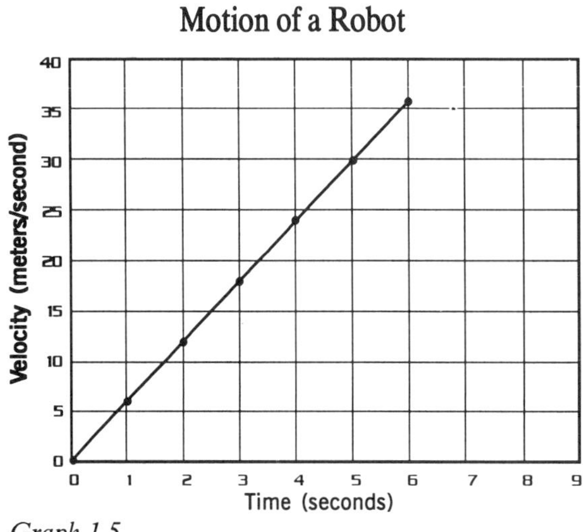

Graph 1.5

After the robot was turned on, it accelerated at a uniform rate from rest to 36.0 m/s in only 6.0 seconds. This period of the robot's motion could be graphed as shown in Graph 1.5.

Notice that this graph looks very much like our position vs. time graph. We can only tell the difference by paying attention to the label on the *y*-axis, which is why teachers tend to deduct credit from students who don't label their graphs properly. The fact that the robot's motion can be described with a straight line here indicates that the acceleration was uniform.

We can find the velocity of the robot at any time given on the graph, by tracing back to the *y*-axis. Can you see, for example, that at 5 seconds the robot was traveling with a velocity of 30 m/s?

To calculate the robot's acceleration during this period of time, we find the slope of the line. I recommend that you choose to compare two points on the line that are very clear. For example, look again at the point on the graph that represents the robot's velocity at 5 seconds. It seems to fall clearly at 30.0 m/s, so that is a good point to use in our comparison. We can also use the origin, which shows us that the velocity of the robot at 0 seconds was 0 m/s.

$$a_{avg} = \text{slope} = \frac{\Delta \text{y axis}}{\Delta \text{x axis}} = \frac{\text{change in velocity}}{\text{change in time}}$$

$$= \frac{30\text{ m/s} - 0\text{ m/s}}{5\text{ s} - 0\text{ s}} = \frac{30\text{ m/s}}{5\text{ s}} = \mathbf{6\ m/s^2}$$

Another interesting thing about these velocity graphs is that we can calculate the displacement of an object by looking at the area *under the plotted line*. Look at Graph 1.5 and notice the area under the line and above the *x*-axis forms a triangle.

If we use the formula for the area of a triangle ($\frac{1}{2}$ base × height), we can find the displacement of the robot during this period of time.

Given: Base = 6 s Height = 36 m/s

Find: Displacement

Solution: Displacement = $\frac{1}{2}$ base × height = 3 $\cancel{s}$ × 36 m/$\cancel{s}$ = **108 m**

So, the robot travels 108 meters (let's ignore the fact that this seems to be an unusually large gym!) during the first 6 seconds. If you doubt our answer, you could always check it with one of our formulas for uniform motion. From the graph we can tell that the robot's initial velocity was 0 m/s, its final velocity was 36 m/s, and the change in time was 6 s. Let's use this information and the acceleration of 6 m/s^2 that we found using the slope formula to solve for displacement.

Given: v_i = 0 m/s v_f = 36 m/s Δt = 6 s a = 6 m/s^2

Find: d

Solution:

$$d = v_i \Delta t + \frac{1}{2} a \Delta t^2 = (0.0\text{ m/s})(6\text{ s}) + (3\text{ m/s}^2)(6\text{ s})^2$$

$$= (3\text{ m/}\cancel{s}^2)(36\ \cancel{s}^2) = \mathbf{108\ m}$$

Next, let's suppose that the robot spends the next two seconds slowing down to a velocity of 25.0 m/s. We can add that data to Graph 1.6 on page 58.

To find the acceleration of the robot from 6 seconds to 8 seconds, we find the slope of the line. The point representing the velocity of the robot at 7 seconds is a little unclear, so let's compare the velocity of the robot at 6 seconds to the velocity of the robot at 8 seconds.

$$a_{avg} = \text{slope} = \frac{\Delta y \text{ axis}}{\Delta x \text{ axis}} = \frac{\text{change in velocity}}{\text{change in time}} = \frac{25 \text{ m/s} - 36 \text{ m/s}}{8 \text{ s} - 6 \text{ s}}$$

$$= \frac{-11 \text{ m/s}}{2 \text{ s}} = \mathbf{-5.5 \text{ m/s}^2}$$

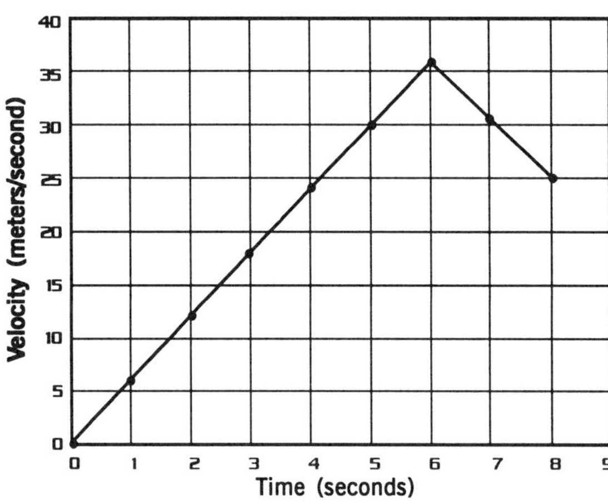

Graph 1.6

This time we ended up with a negative acceleration. Because the velocity prior to this acceleration was positive, this negative acceleration results in a decrease in velocity. As I mentioned before, a negative acceleration doesn't always result in a decrease in velocity, but, in this case, it did.

Once again, you could find the displacement of the robot during this time by finding the area under the graph. If you wanted to find the displacement of the robot from 6 s to 8 s, you could break the area under that part of the graph into one triangle and one rectangle.

We would find the area of the two shapes and add them together to get the total displacement during this time.

$$\text{Area of the triangle} = \frac{1}{2}\text{base} \times \text{height} = 1 \text{ s} \times 11 \text{ m/s} = \mathbf{11 \text{ m}}$$

$$\text{Area of rectangle} = \text{length} \times \text{width} = 25 \text{ m/s} \times 2 \text{ s} = \mathbf{50 \text{ m}}$$

$$\text{Total displacement} = 11 \text{ m} + 50 \text{ m} = \mathbf{61 \text{ m}}$$

Once again, we could check our method with our equation for uniform motion, as shown here:

Given: $v_i = 36\text{ m/s}$ $v_f = 25\text{ m/s}$ $\Delta t = 2\text{ s}$ $a = -5.5\text{ m/s}^2$

Find: d

Solution:

$$d = v_i\Delta t + \frac{1}{2}a\Delta t^2 = (36\text{ m/s})(2\text{ s}) + \left(\frac{1}{2}\right)(-5.5\text{ m/s}^2)(2\text{ s})^2$$
$$= (-2.75\text{ m/}\cancel{s}^2)(4\,\cancel{s}^2) = \mathbf{61\ m}$$

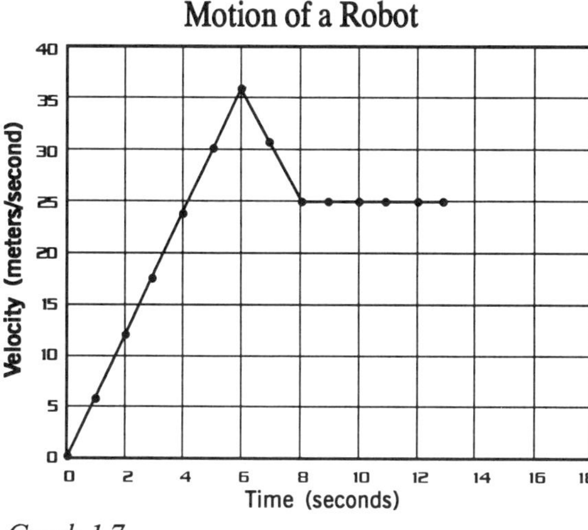

Graph 1.7

Imagine that the robot maintains this velocity for 5 seconds, bringing us to the 13 second mark. Let's add this information to our graph (Graph 1.7).

At a quick glance, you might think that the robot is stationary from 8 seconds to 13 seconds because the slope of the line is zero. However, the robot is still traveling quite fast during this time! It is not the velocity that is zero, it is the acceleration.

$$a_{avg} = \text{slope} = \frac{\Delta y\text{ axis}}{\Delta x\text{ axis}} = \frac{\text{change in velocity}}{\text{change in time}}$$
$$= \frac{25\text{ m/s} - 25\text{ m/s}}{12\text{ s} - 8\text{ s}} = \frac{0\text{ m/s}}{4\text{ s}} = \mathbf{0\ m/s^2}$$

It is not the velocity that is zero, rather, the change in velocity during this period of time is zero. You can tell that the robot is still moving because he is still covering the distance under the curve (25 m/s × 5 s = 125 m) during this time.

Now the robot experiences a constant acceleration of -5.0 m/s^2 for the next 7 seconds. Let's add this data to our graph (Graph 1.8 on page 60).

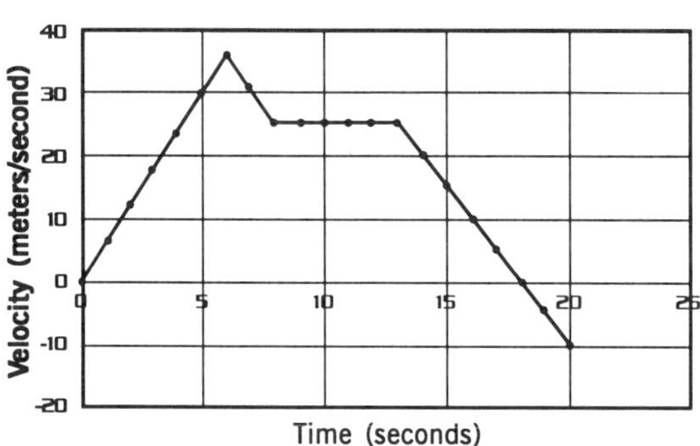

Graph 1.8

What has that crazy robot done this time? It changed directions! Can you see what portion of the graph represents a change in direction? The negative acceleration that the robot experienced caused it to slow down from 13 seconds to 18 seconds. At 18 seconds, the velocity of the robot was actually 0 m/s. Once the line dropped below the origin, you can see that the robot has a negative velocity, so it is heading in the opposite direction to what was considered positive. At this point, the robot is actually speeding up in the negative direction.

Let's summarize what we have learned.

Velocity vs. Time Graphs

1. Uniform acceleration is represented by a straight line segment.
2. For uniform acceleration, the instantaneous acceleration is the same as the average acceleration.
3. The slope of a line segment shows the average acceleration of the object during a period of time.
4. A positive slope represents a positive acceleration.
5. A horizontal line with a slope of zero represents an acceleration of zero.
6. A negative slope represents a negative acceleration.
7. The area under the graphed line represents the displacement.
8. When the line drops below the horizontal axis, the object changes direction.

Lesson 1–6 Review

1. What does the slope of the line on a position vs. time graph represent?
2. What would a negative velocity look like on a position vs. time graph?
3. What would a uniform acceleration look like in a velocity vs. time graph?

Lesson 1–7: Projectile Motion

Most of us became familiar with projectile motion long before we studied physics. Playing catch in the backyard, we observed the motion of a ball as we tossed it to our friend. The ball's path formed a parabola. As the ball came back to us, we saw it rise until it reached its maximum height, and then it started to come down, moving towards us the entire time. The ball would be considered a **projectile** because after you release it, the only force acting upon it (if we ignore air resistance) is gravity. Other common projectiles would include bullets, arrows, snowballs, cannonballs, and basketballs. The motion of such objects is appropriately called **projectile motion**.

Up to this point, we have only considered motion that takes place along one dimension at a time. We talked about objects falling downward, or objects moving forward, but not both at the same time. We know from real-life experiences that objects often do fall downward along what we might call the *y*-axis, while moving forwards along what can be called the *x*-axis, but until now, we weren't ready to study motion along both dimensions at the same time. Now we should be ready. The truth is, if you are careful with your problem-solving, projectile motion problems aren't much harder than the problems we studied in Lesson 1–4.

Horizontal Projectile Motion

The first type of projectile motion that we will consider is often called **horizontal projectile motion**, because all of the projectile's initial velocity will be directed along the horizontal, or *x*-axis. This is the type of motion that you see when an object rolls off the edge of a table, or when a ball is launched parallel to the ground. As you might imagine, an object can have an initial velocity of zero in the vertical (*y*), even when it has a nonzero initial velocity in the horizontal (x). It is also interesting to note that the acceleration in the horizontal (x) dimension is zero, when we ignore air resistance, but the acceleration in the vertical (y) will be equal to the acceleration due to gravity (g), assuming the projectile is launched near the surface of Earth. Let's try an example of a problem and go through the steps involved in the solution.

Example 1

A marble rolls off the edge of a table with a height of 0.755 m and strikes the floor at a distance of 24.3 cm from the edge of the table. Calculate the initial velocity of the ball. (Ignore air resistance.)

These problems are often called two-dimensional motion problems, and we study the motion in each dimension separately. It is a good practice to list the "givens" for the x and y dimensions separately, and label everything carefully. For example, v_{iy} will indicate the initial velocity in the y dimension, and d_x will indicate the displacement in the horizontal dimension.

Convert: The horizontal displacement was given in cm. Let's convert this value to m so that it will match the vertical displacement.

$$24.3\ \cancel{\text{cm}} \times \frac{1\ \text{m}}{100\ \cancel{\text{cm}}} = \mathbf{0.243\ m}$$

Given:

Horizontal → $a_x = 0.0\ \text{m/s}^2$

d_x = how far the ball traveled = 0.243 m

Vertical → $v_{iy} = 0.0\ \text{m/s}$ $a_y = g = -9.81\ \text{m/s}^2$

d_y = height = –0.755 m

Find: v_{ix}

We gave the acceleration in the y (a_y or g) and the displacement in the y (d_y) negative signs because they are both directed downward.

We need to find the initial velocity of the marble in the horizontal. Remember, we define velocity as the rate of displacement, as shown by the formula

$$v_{avg} = \frac{d}{\Delta t}.$$

Because there is no acceleration in the *x* dimension, we can safely assume that the initial horizontal velocity of the marble is the same as the average velocity in the horizontal, so we can solve for our unknown using the formula

$$v_{ix} = \frac{d_x}{\Delta t}.$$

We already know that the displacement in the horizontal dimension is 0.243 m, because that is how far the ball landed from the edge of the table. All that we need to do is to determine the amount of time that the ball was in the air. How do we do that? We know the initial velocity of the ball in the y was zero. We also know the height from which the ball fell, and the acceleration due to gravity, so we simply do a free-fall problem.

Isolate:

Original Formula: $d_y = \frac{1}{2} g \Delta t^2$

Multiply both sides by 2: $2d_y = \cancel{2} \times \frac{1}{\cancel{2}} g \Delta t^2$

Divide both sides by g: $\frac{2d_y}{g} = \frac{\cancel{g} \Delta t^2}{\cancel{g}}$

We get: $\Delta t^2 = \frac{2d_y}{g}$

Now, take the square root of both sides: $\sqrt{\Delta t^2} = \sqrt{\frac{2d_y}{g}}$

We get our working formula: $\Delta t = \sqrt{\frac{2d_y}{g}}$

Solution: $\Delta t = \sqrt{\frac{2d_y}{g}} = \sqrt{\frac{2(-0.243\ \cancel{m})}{-9.81\ \cancel{m}/s^2}} = \mathbf{0.223\ s}$

Armed with both the displacement in the horizontal and the change in time, we can solve for the initial velocity in the horizontal, as shown here:

$$v_{ix} = \frac{d_x}{\Delta t} = \frac{0.243\ m}{0.223\ s} = \mathbf{1.09\ m/s}$$

Notice that I rounded the answer to three significant digits, because the original values in the problem contained three significant digits.

Let's try another example of a horizontal projectile motion, but this time, let's solve for a different unknown. Once again, you should notice that mechanical problem-solving won't be enough to do these types of problems. You must approach each problem that you encounter ready to reason out a solution.

Example 2

An archer stands on the wall of a castle and fires an arrow from a height of 12.10 m above the ground. A level field stretches out in front of the castle wall as far as the eye can see. If the archer fires an arrow parallel to the ground with an initial horizontal velocity of 11.0 m/s, how far will the arrow travel horizontally before hitting the ground? (Ignore air resistance.)

Given:

Horizontal → $a_x = 0.0 \text{ m/s}^2$ $\quad v_{ix} = 11.0 \text{ m/s}$

Vertical → $v_{iy} = 0.0 \text{ m/s}$ $\quad a_y = g = -9.81 \text{ m/s}^2$

$d_y = \text{height} = -12.10 \text{ m}$

Find: d_x

First, we will find out how long the arrow will be in the air using the distance that the arrow must fall to hit the ground with the formula that we found for the previous problem.

Formula: $\Delta t = \sqrt{\dfrac{2d_y}{g}} = \sqrt{\dfrac{2(-12.10 \text{ m})}{-9.81 \text{ m/s}^2}} = \mathbf{1.57\ s}$

Next, we make use of the initial velocity (v_{ix}) in the horizontal and the change in time (Δt) to find the displacement.

Isolate:

Original formula: $v_{ix} = \dfrac{d_x}{\Delta t}$

Multiply both sides by Δt: $\Delta t \times v_{ix} = \dfrac{d_x}{\cancel{\Delta t}} \times \cancel{\Delta t}$

We get: $d_x = v_{ix} \Delta t$

Solution: $d_x = v_{ix} \Delta t = (11.0 \text{ m/}\cancel{\text{s}})(1.57\ \cancel{\text{s}}) = \mathbf{17.3\ m}$

Let's try one more example of a horizontal motion before moving on to the next type of projectile motion problem.

Example 3

An airplane is flying a practice bombing run by dropping bombs on an old shed. The plane is flying horizontally with a speed of 185 m/s. It releases a bomb when it is 593 m away from the shed, and it scores a direct hit. Assuming there is no air resistance, how high was the airplane flying when it dropped the bomb?

You might not recognize this as an example of horizontal projectile motion at first, but the bomb is being dropped while it is moving forward with the same speed as the plane, and it experiences free fall on its way down. You might have noticed that I used the word "speed" instead of "velocity" in this problem, but that it just because I didn't specify the direction or heading of the plane. I could have also said that the magnitude of the velocity is 185 m/s. Either way, we solve the problem the same way.

Given:

Horizontal → $a_x = 0.0\ \text{m/s}^2$ $v_{ix} = 185\ \text{m/s}$ $d_x = 593\ \text{m}$

Vertical → $v_{iy} = 0.0\ \text{m/s}$ $a_y = g = -9.81\ \text{m/s}^2$

Find: d_y

Can you see where you should begin to solve this problem? We will start by using the initial horizontal velocity and the horizontal displacement to determine how long the bomb was in the air, or Δt.

Isolation:

Original formula: $v_{ix} = \dfrac{d_x}{\Delta t}$

Multiply both sides by Δt: $\Delta t \times v_{ix} = \dfrac{d_x}{\cancel{\Delta t}} \times \cancel{\Delta t}$

We get: $\Delta t \times v_{ix} = d_x$

Divide both sides by v_{ix}: $\dfrac{\Delta t \cancel{v_{ix}}}{\cancel{v_{ix}}} = \dfrac{d_x}{v_{ix}}$

We get: $\Delta t = \dfrac{d_x}{v_{ix}}$

Solution: $\Delta t = \dfrac{d_x}{v_{ix}} = \dfrac{593\,\text{m}}{185\,\text{m/s}} = \mathbf{3.21\,s}$

Now that we know how long it takes the bomb to fall, we can calculate the bomb's vertical displacement, as shown below.

Formula: $d_y = \dfrac{1}{2} g\Delta t^2 = (0.5)(-9.81\,\text{m/s}^2)(3.21\,\text{s})^2 = \mathbf{-50.5\,m}$

The negative sign simply indicates that the bomb fell downward. We won't include the negative sign in our final answer, because we were solving for the altitude of the plane, not the displacement of the bomb.

Answer: The plane was flying at a height of **50.5 m**

Parabolic Motion

Let's move on to the other major type of projectile motion problem, which is sometimes called a **parabolic motion** problem because of the shape of the paths these objects take. In this type of problem, a projectile is given an initial velocity in two dimensions, so we need to go over a technique for dealing with our initial velocity vectors. Our first task in this type of problem will often be to take the initial velocity vector and resolve it, or break it up, into horizontal and vertical components. Once we have the component velocity vectors, we can add them to our list of "givens."

Resolution of Vectors

Breaking an individual vector up into two or more component vectors.

By now, you should be getting the hang of vector addition, so I would like to add another "tool" to your mental "toolbox." To do parabolic motion problems, you need to be able to break a vector apart into its individual components, rather than adding components to find a resultant vector. Let's imagine that a cannon is set up so the barrel points at an angle of 34.0° above the horizontal. When the cannon is fired, the ball launches with an initial velocity of 23.0 m/s at an angle of 34.0° above the ground. How could we figure out how fast the ball is moving upward, and how fast it is moving forward as it leaves the barrel of the cannon?

Draw a horizontal line to represent the ground, and add the initial velocity vector. Unless you plan on drawing everything to scale and using a protractor and a ruler to solve graphically, the diagram doesn't need to be drawn to scale.

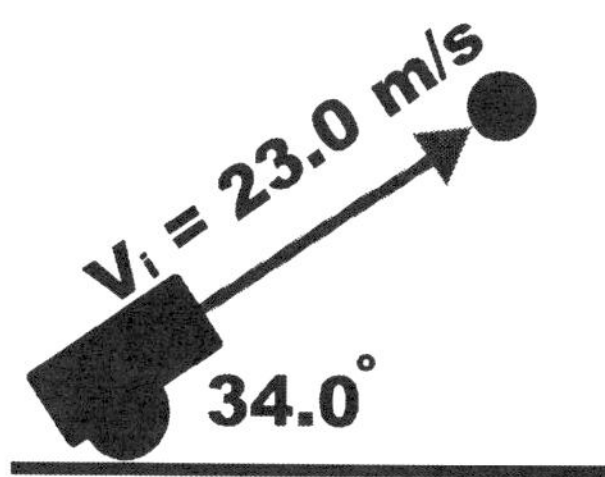

Figure 1.7

Now, if we draw a vertical line from the tip of the velocity arrow to the horizontal line representing the ground, so that this new line is perpendicular to the horizontal, we have a right triangle.

Notice how the initial velocity vector forms the hypotenuse of our right triangle. Turn the other sides of the triangle into arrows to represent the horizontal (x) and vertical (y) components of our initial velocity vector.

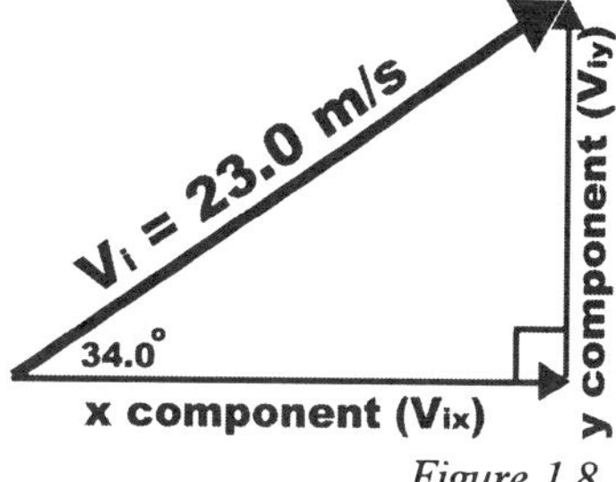

Figure 1.8

Remember that this is a right triangle, and we know all three angles and one side. This means that we can use trigonometry functions to find the magnitude of the remaining two sides.

Trigonometry Functions

$$\sin = \frac{\text{opp}}{\text{hyp}} \qquad \cos = \frac{\text{adj}}{\text{hyp}} \qquad \tan = \frac{\text{opp}}{\text{adj}}$$

v_{iy} = opposite side
= hypotenuse × sin θ
= (23.0 m/s)(sin 34.0°) = **12.9 m/s**

v_{ix} = adjacent side
= hypotenuse × cosθ
= (23.0 m/s)(cos 34.0°) = **19.1 m/s**

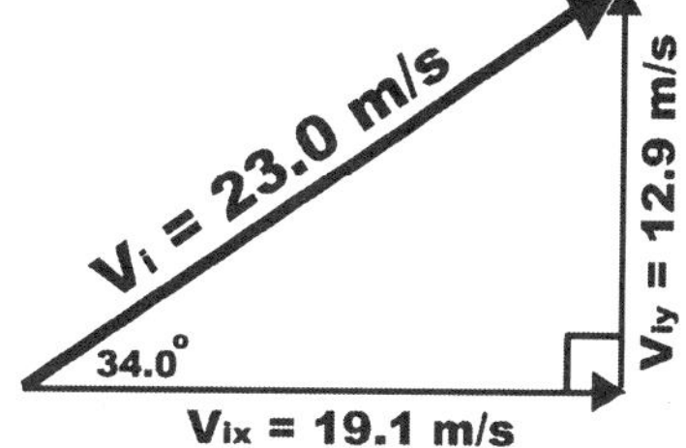

Figure 1.9

Now that we have the horizontal (x) and vertical (y) components of the initial velocity, we can use this information to figure out how long the ball will be in the air (Δt), how high the ball will go (max d_y), and how far the ball will travel (d_x). Let's put an entire problem together in a format similar to what you are likely to encounter.

Example 4

A boy kicks a soccer ball, giving it an initial velocity of 7.50 m/s at an angle of 27.0° above the horizontal. How high will the ball go? How long will it be in the air? How far will it travel?

Before we even list our givens, let's resolve the original velocity vector into its x and y components, just as we did with the cannonball.

Resolution:

$v_{iy} = v_i \sin\theta = (7.50 \text{ m/s})(\sin 27.0°) = 3.40 \text{ m/s}$

$v_{ix} = v_i \cos\theta = (7.50 \text{ m/s})(\cos 27.0°) = 6.68 \text{ m/s}$

I also want to remind you of an "implied given" that we can make use of in this problem. The soccer ball will slow down as it goes upward until its velocity reaches zero at its maximum height (max d_y). The ball will then fall downward, so the velocity at this maximum height is not really its final velocity, but it is the final velocity for the upward part of its journey. We can use this concept to help us find out how high the ball will go, and how long it will be in the air.

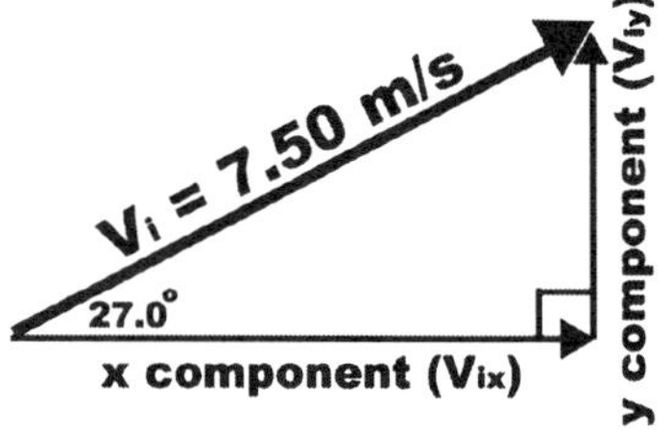

Figure 1.10

Given: Horizontal → $v_{ix} = 6.68$ m/s $a_x = 0$ m/s

Vertical → $v_{iy} = 3.40$ m/s $a_y = -9.81$ m/s2

v_{fy} (at max d_y) = 0 m/s

Find: Maximum height (d_y) Time of flight (Δt)

distance traveled (d_x)

Maximum Height

Isolation:

Starting with the formula: $v_f^2 = v_i^2 + 2gd$

We modify it to work with the y dimension:

$v_{fy}^2 = v_{iy}^2 + 2gd_y$ v_{fy} (at max d_y) = 0,

so we cross that out: $\cancel{v_{fy}^2} = v_{iy}^2 + 2gd_y$

Isolating d_y, we get: $d_y = \frac{-v_{iy}^2}{2g}$

Solution: $d_y = \frac{-v_{iy}^2}{2g} = \frac{-(3.40\text{ m/s})^2}{2(-9.81\text{ m/s}^2)} = \mathbf{0.589\text{ m}}$

Time of Flight

We will figure out how long it takes the ball to travel to its maximum height and then double that change in time, because the ball will spend an equal amount of time falling from that height.

Isolate:

Starting with the formula: $v_f = v_i + g\Delta t$

We modify it for the y dimension: $v_{fy} = v_{iy} + g\Delta t$

v_{fy} (at max d_y) = 0, so we cross that out: $\cancel{v_{fy}} = v_{iy} + g\Delta t$

Isolating Δt, we get: $\Delta t = \frac{-v_{iy}}{g}$

Solution: $\Delta t = \frac{-v_{iy}}{g} = \frac{-3.40\,\cancel{\text{m/s}}}{-9.81\,\cancel{\text{m/s}^2}} = 0.347\text{ s}$

But remember, this represents the change in time for the upward motion of the ball only! In order to find the total change, we must multiply this value by 2.

Total $\Delta t = 2 \times 0.347\text{ s} = \mathbf{0.694\text{ s}}$

Distance Traveled

Solution: $d_x = v_{ix}\Delta t = (6.68\text{ m/}\cancel{\text{s}})(0.694\,\cancel{\text{s}}) = \mathbf{4.64\text{ m}}$

Lesson 1–7 Review

1. When we do projectile motion problems, what value do we use for acceleration in the *x* dimension?
2. If one bullet was fired horizontally from a height of 2.00 m and another was dropped from the same height at exactly the same time, which would hit the ground first? (Ignore air resistance, and assume a level horizontal field for the first bullet to travel along until it fell to the ground.)
3. If a football is kicked with an initial velocity of 12.0 m/s at an angle of 33.0° above the horizontal, what is the initial velocity in the *y* dimension?

Chapter 1 Examination

Part I—Matching

Match the following terms to the definitions that follow.

a. displacement	d. free fall	g. mechanics
b. kinematics	e. scalar	h. vector
c. velocity	f. acceleration	i. component vector

_____1. A quantity that can be described completely with numbers and units.

_____2. The branch of physics that deals with motion.

_____3. Speed in a particular direction.

_____4. One of a set of two or more vectors that can combine to form a resultant vector.

_____5. The rate at which an object changes its velocity.

_____6. The state of a falling object where the only unbalanced force acting on it is gravity.

Part II—Multiple Choice

For each of the following questions, select the best answer.

7. Which of the following is not an example of a vector quantity?

   a) distance c) velocity
   b) displacement d) acceleration

8. Which of the following shows the correct units for acceleration?

   a) m b) m/s c) m/s^2 d) m^2/s^2

9. When an object with a velocity of zero experiences a negative acceleration, it will

   a) remain at rest c) slow down
   b) speed up d) stop

10. A dog travels 3.0 m east, 6.0 m north, and 2.0 m south. What total distance did the dog travel?

    a) 11.0 m c) 9.0 m
    b) 6.0 m to the northeast d) 5.0 m

11. A dog walks 2.0 m north, 5.0 m south, and then 3.0 m north. What is the dog's overall displacement?

    a) 10.0 m b) 0.0 m c) 7.0 m south d) 2.0 m east

12. An object in motion must experience a change in

    a) speed b) acceleration c) velocity d) position

13. If a boy dropped a rock off a high cliff, how fast would it be falling after 2.0 seconds had elapsed? (Ignore air resistance.)

    a) -2.0×10^1 m/s c) 35 m/s
    b) 2.0×10^2 m/s d) -35 m/s

14. If an apple fell from a tree, how long would it take to reach a velocity of -3.00 m/s?

    a) 3.00 s b) 3.27 s c) 2.94 s d) 0.306 s

15. A car starts from rest and accelerates at a constant rate of 4.50 m/s^2 until it reaches a velocity of 26.0 m/s. How far does it travel during this time?

    a) 5.78 m b) 21.5 m c) 75.1 m d) 94.8 m

Part III—Graphs

Base your answers to questions 16 through 18 on the Graph 1.9.

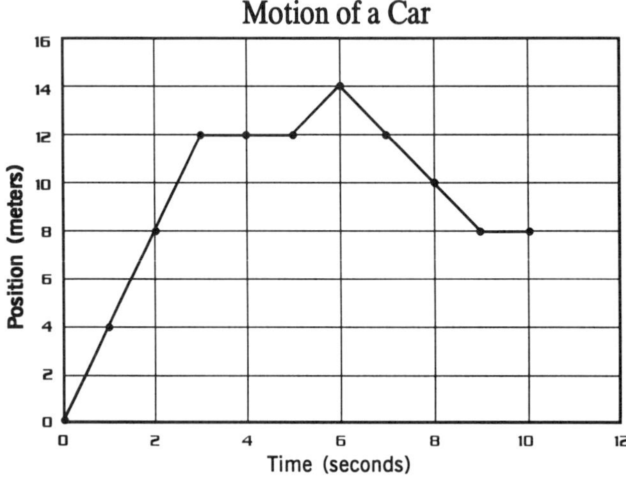

Graph 1.9

16. How fast is the car going at 2 seconds?
17. How far does the car travel in the first 3 seconds?
18. At what time does the car first stop?
19. During which time interval does the car have a negative velocity?
20. At 10 s, how far away is the car?

Part IV—Calculations

Perform the following calculations.

21. An apple with an initial velocity of 1.5 m/s rolls horizontally off the edge of a table with a height of 55 cm. How long will the apple be in the air, and how far from the edge of the table will it strike the ground? (Ignore air resistance.)
22. A car launches horizontally off the edge of a cliff with a height of 18.6 m. It strikes the ground below at a distance of 98.4 m away from the edge of the cliff. How fast was the car going when it flew off the edge of the cliff?

23. A golfer hits a golf ball with an initial velocity of 11.4 m/s at an angle of 43.0° above the horizontal. Assuming the ground in front of the golfer is level and there is no air resistance, how far will the ball travel?

24. A truck with an initial velocity of 35.0 m/s experiences a constant acceleration of $-6.5\ m/s^2$ as the driver applies the brakes. How long does it take the truck to come to a complete stop?

25. A boy drops a water balloon off the top of his apartment building and counts how long it takes to hit the ground below. If he is able to count to "3 Mississippis" (approximately 3 seconds) before the balloon hits the ground, approximately (to one significant digit) how tall is the building?

Answer Key

The actual answers will be shown in brackets, followed by an explanation. If you don't understand an explanation that is given in this section, you may want to go back and review the lesson that the question came from.

Lesson 1–1 Review

1. [displacement]—You can't completely describe an object's displacement without indicating the direction.
2. [98 m]—The direction that she traveled is not important, because the question only asks us for the distance that she traveled (33 m + 42 m + 23 m = 98 m).
3. [36 m at 66° north of east]—This question asks for a displacement, so we must indicate the change of position and the direction. We can find the overall change in position using the Pythagorean theorem:

$$d_r = \sqrt{{d_1}^2 + {d_2}^2} = \sqrt{(15\text{ m})^2 + (33\text{ m})^2} = 36.249\text{ m, which we will round to }36\text{ m}.$$

To find the direction, we calculate the angle formed between the resultant vector and the displacement to the east.

$$\theta = \tan^{-1}\frac{\text{opposite side}}{\text{adjacent side}} = \tan^{-1}\frac{33\ \cancel{\text{m}}}{15\ \cancel{\text{m}}} = 66°\text{ north of east}$$

Lesson 1–2 Review

1. [velocity]—Remember, *v*elocity is a *v*ector. *S*peed is *s*calar.
2. [yes]—When an object changes direction, the velocity will change, even if the speed is constant.
3. [130 m/s at 16° west of south]—We can find the magnitude of the velocity using the Pythagorean theorem,

   $$v_r = \sqrt{v_p^2 + v_w^2} = \sqrt{(125\text{ m/s})^2 + (35\text{ m/s})^2} = 130\text{ m/s},$$

   and the direction with the inverse tangent,

   $$\theta = \tan^{-1}\frac{\text{opposite side}}{\text{adjacent side}} = \tan^{-1}\frac{35\text{ m/s}}{125\text{ m/s}} = 16°\text{ west of south}$$

Lesson 1–3 Review

1. [acceleration]—$\text{acceleration} = \dfrac{\text{change in velocity}}{\text{change in time}}$
2. [13 s]—$\Delta t = \dfrac{v_f - v_i}{a} = \dfrac{125\text{ m/s} - 0.0\text{ m/s}}{9.5\text{ m/s}^2} = 13\text{ s}$
3. [yes]—If an object with a negative velocity experiences a negative acceleration, it speeds up. The negative sign simply implies a direction.

Lesson 1–4 Review

1. [17 m/s]—$v_f = a\Delta t = (4.5\text{ m/s}^2)(3.8\text{ s}) = 17\text{ m/s}$
2. [18.4 m/s²]—$a = \dfrac{v_f^2}{2d} = \dfrac{(45.0\text{ m/s})^2}{2(55.0\text{ m})} = 18.4\text{ m/s}^2$
3. [75.0 m]—
   $$d = \left(\frac{v_i + v_f}{2}\right)\Delta t = \left(\frac{5.00\text{ m/s} + 25.0\text{ m/s}}{2}\right)5.00\text{ s}$$
   $$= (15.0\text{ m/s})(5.00\text{ s}) = 75.0\text{ m}$$

   Note, the 2 in the denominator of our equation is considered a *counting number*, not a measurement, so we don't use it for determining the number of significant digits in our answer.

Lesson 1–5 Review

1. [free fall]—True free fall only occurs when the only unbalanced force acting on it is gravity.

2. [The velocity will equal zero, the acceleration will equal g.]—Remember that the velocity decreases as the ball goes up, until it reaches zero at the top of its flight. The velocity will then increase on the way down. The acceleration is constant. During the entire flight, the ball experiences the acceleration due to gravity, $-9.81\ \text{m/s}^2$.

3. [326 m]—$d = \frac{v_f^{\ 2} - v_i^{\ 2}}{2g} = \frac{(0.0\ \text{m/s})^2 - (80.0\ \text{m/s})^2}{2(-9.81\ \text{m/s}^2)} = 326\ \text{m}$

Lesson 1–6 Review

1. [velocity]—Remember, we find the slope of the line by dividing the change in the *y*-axis by the change in the *x*-axis. Because the *y*-axis on this type of graph represents position and the *x*-axis represents time, we are really finding the displacement over the change in time, or velocity, when we take the slope.
2. [a line segment with a negative slope]—Remember to always subtract the initial position from the final position ($\Delta y = y_f - y_i$) when using the slope formula. This way, you get a negative value for the change in position when you should.
3. [a straight line segment]—Only a horizontal line represents an acceleration of zero, but any straight line segment shows uniform acceleration.

Lesson 1–7 Review

1. [zero]—We ignore air resistance, so the only unbalanced force acting on the object, after the initial projection, is gravity. In other words, the projectile is in free fall.
2. [neither. They would hit the ground at the same time.]—The first bullet only appears to be in the air longer because it travels a great distance in the *y* dimension. Remember, both bullets are in free fall, with an initial velocity of zero in the *x* dimension. The amount of time that each bullet will be in the air can be found with the same calculation.

$$\Delta t = \sqrt{\frac{2d_y}{g}} = \sqrt{\frac{2(-2.00\ \cancel{\text{m}})}{-9.81\ \cancel{\text{m}}/\text{s}^2}} = 0.639\ \text{s}$$

3. [6.54 m/s]—$v_{iy} = v_i \sin\theta = (12.0\ \text{m/s})(\sin 33.0^\circ) = 6.54\ \text{m/s}$

Chapter 1 Examination

1. [e. scalar]—A scalar quantity, such as speed or distance, requires that no direction be specified.

2. [g. mechanics]—Classical, or Newtonian, mechanics deals with the motion of relatively large objects, while quantum mechanics deals with the motion of particles.
3. [c. velocity]—Velocity is the vector counterpart to speed.
4. [i. component vector]—Remember, *component* vectors *combine* to form resultant vectors.
5. [f. acceleration]—Remember, unlike the common definition for acceleration, in physics, an object can accelerate without changing its speed, because a change in direction is also a change in velocity.
6. [d. free fall]—An object such as a parachute, that is being slowed by significant air resistance is not in free fall.
7. [a. distance]—You can describe the distance you walked without mentioning a direction. A person may say, "I ran 5 miles today!" and you know what they mean.
8. [c. m/s^2]—Velocity is measured in m/s, and a change in time is measured in seconds.

   If $\text{acceleration} = \dfrac{\text{velocity}}{\text{change in time}}$ then it must be measured in $\dfrac{\text{m/s}}{\text{s}}$, or m/s^2.

9. [b. speed up]—An acceleration is a change in velocity, so answers A and D don't make sense. Answer C makes no sense because how can an object at rest slow down? The object will speed up, increasing its velocity in the direction that has been designated as negative.
10. [a. 11.0 m]—To find the total distance, we simply add all of the individual distances: 3.0 m + 6.0 m + 2.0 m = 11.0 m.
11. [b. 0.0 m]—The dog ends up in exactly the same spot, so its displacement is zero.
12. [d. position]—An object at constant speed or velocity is still in motion, provided the speed is not zero. An object with an acceleration of zero can certainly be in motion. Only the position of an object in motion needs to change.
13. [a. -2.0×10^1 m/s]—
   $$\begin{aligned} v_f &= v_i + g\Delta t = 0.0\text{ m/s} + (-9.81\text{ m/s}^2)(2.0\text{ s}) \\ &= -2.0 \times 10^1\text{ m/s} \end{aligned}$$
   Remember, the negative sign indicates that the velocity is downward.
14. [d. 0.306 s]—$\Delta t = \dfrac{v_f}{g} = \dfrac{-3.00\text{ m/s}}{-9.81\text{ m/s}^2} = 0.306\text{ s}$

15. [c. 75.1 m]—$d = \frac{v_f^2 - v_i^2}{2a} = \frac{(26.0\text{ m/s})^2 - (0.0\text{ m/s})^2}{2(4.50\text{ m/s}^2)} = 75.1\text{ m}$

16. [4 m/s]—You must find the slope of the line at that point. You can select any two points from the straight-line segment and use the slope formula.

$$\text{velocity} = \text{slope} = \frac{\Delta y\text{ axis}}{\Delta x\text{ axis}} = \frac{8\text{ m} - 0\text{ m}}{2\text{ s} - 0\text{ s}} = 4\text{ m/s}$$

17. [12 m]—We can look at the *y*-axis and see that the car went from 0 m to 12 m during this period of time.

18. [3 s]—A slope of zero represents a velocity of zero. The slope becomes zero at 3 s.

19. [6 s to 9 s]—From 6 seconds to 9 seconds, the slope of the line is negative, as shown by the slope formula.

$$\text{velocity} = \text{slope} = \frac{\Delta y\text{ axis}}{\Delta x\text{ axis}} = \frac{8\text{ m} - 14\text{ m}}{9\text{ s} - 6\text{ s}} = -2\text{ m/s}$$

20. [8 m]—Tracing back to the *y* axis, we can see that the car has a displacement of 8 m.

21. [0.33 s, 0.50 m]—First, convert the height of the table to meters.

$$55\ \cancel{\text{cm}} \times \frac{1\text{ m}}{100\ \cancel{\text{cm}}} = 0.55\text{ m}$$

Given: Horizontal → v_{ix} = 1.5 m/s a_x = 0.0 m/s^2
Vertical → v_{iy} = 0.0 m/s d_y = h = −0.55 m a_y = g = −9.81 m/s^2

Solution: $\Delta t = \sqrt{\frac{2d_y}{g}} = \sqrt{\frac{2(-0.55\text{ m})}{-9.81\text{ m/s}^2}} = 0.33\text{ s}$

$d_x = v_{ix}\Delta t = (1.5\text{ m}/\cancel{\text{s}})(0.33\ \cancel{\text{s}}) = 0.495\text{ m}$, which rounds to 0.50 m.

22. [50.5 m/s]—Given: Horizontal → a_x = 0.0 m/s^2 d_x = 98.4 m
Vertical → d_y = −18.6 m a_y = g = −9.81 m/s^2

First, let's figure out how long the car was in the air.

$$\Delta t = \sqrt{\frac{2d_y}{g}} = \sqrt{\frac{2(-18.6\text{ m})}{-9.81\text{ m/s}^2}} = 1.95\text{ s}$$

It traveled 98.4 m in this time, so its initial velocity was

$$v_{ix} = \frac{98.4\text{ m}}{1.95\text{ s}} = 50.5\text{ m/s}$$

23. [13.2 m]—First, we will resolve the initial velocity into its x and y components.

$$v_{ix} = v_i \cos\theta = (11.4 \text{ m/s})(\cos 43.0°) = 8.34 \text{ m/s}$$

$$v_{iy} = v_i \sin\theta = (11.4 \text{ m/s})(\sin 43.0°) = 7.77 \text{ m/s}$$

Given: Horizontal → $v_{ix} = 8.34$ m/s $\quad a_x = 0.0 \text{ m/s}^2$

Vertical → $v_{iy} = 7.77$ m/s $\quad ay = g = -9.81 \text{ m/s}^2$

v_{fy} (at max d_y) = 0 m/s

Now, we use the v_{iy} to find out how long the ball spends going up.

$$\Delta t = \frac{v_{fy} - v_{iy}}{g} = \frac{0 \text{ m/s} - 7.77 \text{ m/s}}{-9.81 \text{ m/s}^2} = 0.792 \text{ s}$$

The ball spends an equal amount of time coming down, so the total time of flight is 2 × 0.792 s = 1.58 s.

Finally, we use this time of flight and the velocity in the horizontal to determine how far the ball flies. $d_x = v_i \times \Delta t$ = (8.34 m/s)(1.58 s) = 13.2 m

24. [5.4 s]—$\Delta t = \dfrac{v_f - v_i}{a} = \dfrac{0.0 \text{ m/s} - 35.0 \text{ m/s}}{-6.5 \text{ m/s}^2} = 5.4 \text{ s}$

25. [40 m]—$d = \dfrac{1}{2} g \Delta t^2 = (0.5)(-9.81 \text{ m/s}^2)(3 \text{ s})^2 = -44.145 \text{ m} = 40 \text{ m}$

The answer to our calculation was negative because we solved for the displacement of the balloon, which fell approximately 40 meters in the downward direction. However, the question asked about the height of the building, so we answer with the absolute value.

Forces and the Laws of Motion

In this chapter we will begin to study forces and the effects that forces have on objects. A force is simply a push or a pull. You exert a force on a door when you push it open or pull it closed. Earth exerts a force on the moon, allowing it to stay in orbit. Sometimes your sweater might exert a force on a sock after you take them out of the dryer, resulting in what is called "static cling." Forces exert an influence on objects, sometimes causing them to move and sometimes causing them to stop. Let's explore these forces and motion in more detail.

Lesson 2–1: Forces

We spend our whole lives living under the influence of various forces. The force of gravity exerted by Earth keeps us on the ground. The electrostatic force of repulsion between our electrons and the floor's electrons keeps us from falling through it. The force of friction between our shoes and the floor allows us to move forward. A strong nuclear force keeps the subatomic particles in the nuclei of our atoms from flying apart. You exert a force on a fork to move it towards your mouth, so your teeth can exert forces on some food. Get the idea?

Despite being surrounded by forces, many people have a hard time defining or describing forces. In simplest terms, we describe a **force** as a push or a pull. Alternatively, we can think of a force as something that is capable of changing the velocity of an object. Remember, the velocity of an object includes both speed and direction, so a force can change either one or the other, or both.

Force—A push or a pull. Something that is capable of changing an object's velocity.

Some forces are called **contact forces**. Contact forces require contact between two objects. We need to touch a door to push it open, so that is an example of a contact force. Other forces, called **field forces**, don't require physical contact to work. Field forces, which include gravity and magnetism, act over distances. The sun is so far away from Earth that it takes light, which can travel at 3.00×10^8 m/s, about 8 minutes to get here from there. However, even across that incredible distance, some 1.5×10^8 km (or 9.3×10^7 miles), the force of gravity between the sun and Earth keeps our planet in orbit! And yet, physicists consider gravity a relatively weak force!

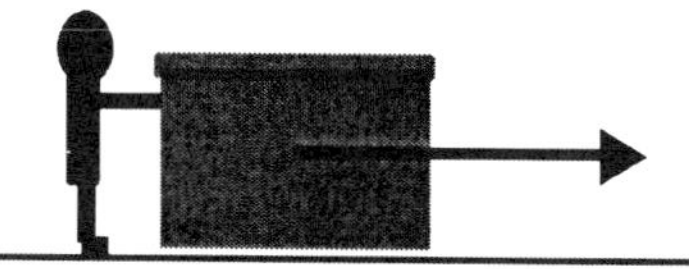

Figure 2.1

Forces are vector quantities, which means that a direction is required to describe them completely. At any given moment, an object can have several forces exerted on it. Imagine, for example, a student is pushing on a desk (exerting a force on it), causing it to slide (change its velocity) across the floor.

There are a number of forces acting on the desk. Because each force is a vector, all of the forces can be added together, using the techniques described in chapter one, to find the resultant vector. This resultant force vector is sometimes referred to as the net force (F_{net}) acting on the object.

Net Force (F_{net})—The vector sum of all of the forces acting on an object.

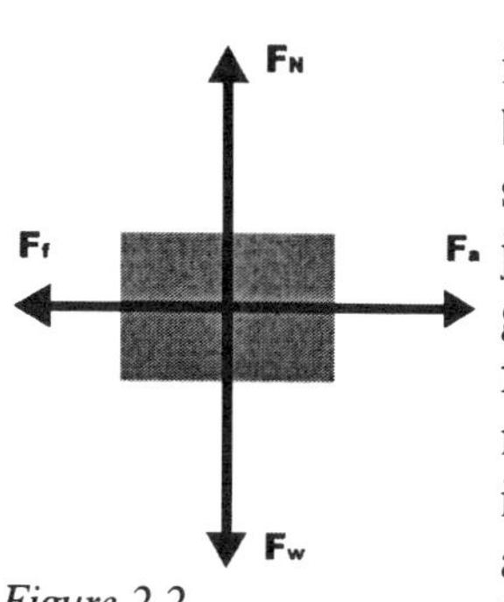

Figure 2.2

When you want to determine the net force acting on an object, it is often useful to draw a **free-body diagram**. A free-body diagram is a simple sketch that highlights the forces acting on an object. The object can be represented with a simple geometric shape, such as a square or circle. The forces acting on the object are represented by arrows. Each arrow points in the direction of the force it represents, and the length of the arrow can give an indication of the relative magnitude of the force. However, it is not always practical to draw the vectors to scale, as some forces involved may be much larger than others. Examine the free-body diagram of the desk with the student pushing on it.

Do you recognize what each of the arrows shown in Figure 2.2 represent? The **applied force** (f_a) is the force being exerted by the student on the desk. The weight (f_w) is the force exerted by Earth on the desk. The normal force (f_N) is the force being exerted by the floor on the desk perpendicular to the surface of the floor, and, as you can surmise from the length of the respective arrows, it is equal to the magnitude of but opposite in direction to the force of weight. The friction (f_f) is the force exerted on the desk by the floor in the direction opposite to its motion, and, judging by the length of the respective arrows, it is less than the applied force and in the opposite direction.

Notice that the free-body diagram only shows the forces acting on the desk, because that is the object that we are interested in. We don't include the forces the desk is exerting on the student or the floor, because that is not what we want to focus on.

Let's try a problem where we are asked to draw a free-body diagram and solve for the net force being exerted on a desk.

Example 1

A student applies a force of 80.0 N to a desk with a weight of 700.0 N in a direction that is horizontal to the surface of the floor. The force of friction between the desk and the floor is 60.0 N. Draw a free-body diagram and find the net force acting on the desk.

We can sketch the desk as a simple rectangle and represent each of the forces with an arrow. If we wanted to draw the forces to scale, we could make up a scale. For example, we could say 20.0 N = 1.00 cm, and draw the vectors to scale. However, we will be adding the vectors algebraically, so I won't be drawing my diagram to scale.

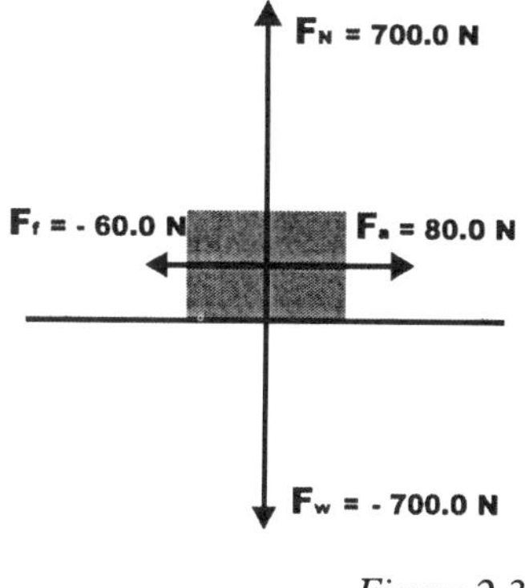

Figure 2.3

As far as the forces go, the problem only mentioned three. There is no mention of the normal force exerted upwards by the floor on the desk. Problems often fail to mention that force because they expect students to realize that the force is there, and that, in situations like the one described here, it is always equal to the weight in magnitude. The reason for this will be explained in Lesson 2–4. If the normal force wasn't there, or if it was less than the weight of the desk, the desk would fall through the floor! Always be sure to add the normal force to this type of problem.

Now let's calculate the net force acting on the desk. Let's designate upwards as positive and downwards as negative. Let's also make the direction of the applied force positive and the direction of the force of friction negative.

Given: F_a = 80.0 N F_f = -60.0 N F_w = 700.0 N F_N = -700.0 N

Find: F_{net}

We need to analyze the horizontal and vertical forces independently.

Solution: $F_{nety} = F_w + F_N = 700.0\text{ N} + (-700.0\text{ N}) = \mathbf{0\ N}$

$$F_{netx} = F_a + F_f = 80.0\text{ N} + (-60.0\text{ N}) = \mathbf{20.0\ N}$$

We have a net force of 20.0 N horizontal to the surface of the floor.

Of course, force vectors can be added together even if they don't lie along the same axis, using the methods we showed for velocity vectors in Lesson 1–1. Try the following example on your own, and then check my answer.

Example 2

Two students attempt to push a box by exerting forces of 28.0 N and 35.8 N, respectively, at right angles to each other, as shown in the free-body diagram in Figure 2.4. Find the resultant force applied by the two students.

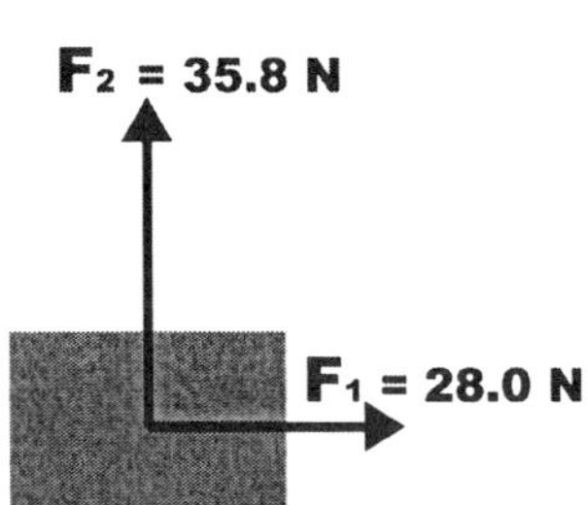

Figure 2.4

Given: F_1 = 28.0 N F_2 = 35.8 N

Find: Resultant force (F_r)

Let's take our two component forces and make a triangle with them and the resultant force.

Solution: We can find the magnitude of the resultant force using the Pythagorean theorem.

$$F_r = \sqrt{F_1^{\,2} + F_2^{\,2}} = \sqrt{(28.0\text{ N})^2 + (35.8\text{ N})^2} = \mathbf{45.4\ N}$$

We find the direction of the force using $\tan^{-1}$.

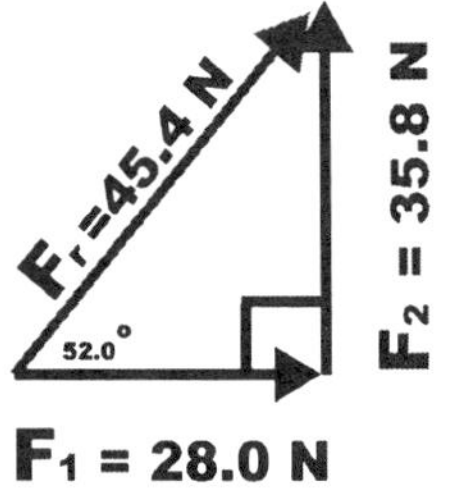

Figure 2.5

$$\theta = \tan^{-1}\frac{\text{opposite side}}{\text{adjacent side}} = \tan^{-1}\frac{35.8\ \cancel{N}}{28.0\ \cancel{N}} = \mathbf{52.0°}$$

Our resultant force is **45.4 N at 52.0° west of south.**

So, now you should understand what a force *is*. In the next few lessons you will be learning more specifically what a force *does*.

Try the following review questions before moving on to the next lesson.

Lesson 2–1 Review

1. _______________ is the vector sum of all of the forces acting on an object.
2. If a boy applies a horizontal force of 12.0 N to push a book across a table, where the force of friction between the book and table is 6.5 N, what will be the net force on the book?
3. If a force of 5.00 N acts concurrently (at the same time) with a 8.00 N force on the same object, at right angles to each other, what will be the magnitude of the resultant force?

Lesson 2–2: Newton's First Law of Motion

Imagine going out to a park, and rolling a basketball down a slide. How far would the ball roll after hitting the ground? Doesn't that depend upon the surface of the ground that the ball hits? Would it go very far if it landed in sand? What if the ground at the bottom of the slide were covered in grass? What if it were cement? How about if it were ice? It was this type of "thought experiment" that lead Galileo Galilei to come up with the concept of **inertia**, paving the way for Sir Isaac Newton's first law of motion.

Before Galileo, the "physics" taught in schools was based on Aristotle's teachings on motion. Aristotle believed that when an object on Earth had no net force acting on it, it would eventually come to a stop, because "rest" was the natural state for the object. Galileo disagreed. His experiments, both mental and actual, lead him to believe that it was just as natural for an object in motion to stay in motion as it was for an object at rest to stay at rest, unless acted upon by an unbalanced or net force.

If you take the example of the ball rolling down the slide, you can imagine that the ball will travel further on a smoother surface. So, if the ball rolled down the slide onto a smooth sheet of ice it would travel farther than if it landed on grass or gravel. Galileo concluded that if an object was moving on a perfectly smooth (frictionless) surface, it would remain in motion unless something (an unbalanced force) caused it to stop. Galileo recognized friction as an unbalanced force that caused moving objects to come to a stop after you stopped applying a force to them.

If friction is what causes objects to stop, what causes them to keep on moving, even after we stop applying forces to them? That would be inertia. Inertia is an object's resistance to any changes in its motion. When a hockey player slaps a puck, the force exerted by his stick sets the puck in motion, but the puck's own inertia allows it to remain in motion until it is acted upon by another force, say, another stick.

There are many real-life examples where you have experienced inertia. Have you ever been in a vehicle, a car or a roller coaster perhaps, that made a sharp turn? You probably felt yourself slide towards the outside of the turn, causing you to be pressed against the person or door to your side. It was your body's own inertia that was resisting the change in direction.

Travel coffee mugs are popular items, mainly because of inertia. Have you ever tried to drink a beverage in a cup with no lid in a moving car? When the driver steps on the gas, the beverage sloshes over the front of your cup and lands in your lap. When the driver steps on the brakes, the liquid sloshes over the back of your cup and spills on your knee. When the driver makes a quick right, the drinks spills over the left side of the cup and lands on your shoe. It is no fun. The inertia of the liquid causes it to resist the changes in motion.

An interesting fact is that an object still possesses inertia even when it is weightless. This means that under weightless conditions, you could still tell the difference between two weightless objects that looked identical, but had different masses. The more massive object would still take more force to accelerate than the less massive object. They have instruments called inertial balances that can be used to mass objects by their inertia.

Galileo noted that an object in motion stays in motion unless acted upon by an unbalanced force, and an object at rest remains at rest unless acted upon by an unbalanced force. Sir Isaac Newton, who was born in 1643, the year after Galileo died, took the idea of inertia as his first law of motion.

Newton's First Law: The Law of Inertia

In the absence of an unbalanced (net) force, an object in motion will remain in uniform motion, and an object at rest will remain at rest.

Newton, who was apparently not usually known for his modesty, once wrote in a letter to fellow scientist, Robert Hooke, "If I have seen further than others, it is by standing upon the shoulders of giants." The choice of the word "giants" may have been a thinly veiled insult aimed at Hooke, who apparently was short in stature, but I like to imagine that he might have been acknowledging the work of Galileo, who deserves much of the credit for Newton's law of inertia.

Lesson 2–2 Review

1. ______________ is an object's tendency to resist changes in motion.

Lesson 2–3: Newton's Second Law of Motion

One of the implications of the law of inertia is that no force is necessary to *keep* an object in uniform motion. A spaceship crossing the vast space between galaxies would not need to burn fuel to maintain constant velocity. It's own inertia is all the ship would require to continue moving at a steady speed in the same direction. Expending fuel would be necessary, however, if the pilot of the ship wanted to speed up, slow down, or change directions. Why is that the case? Newton's second law of motion!

What do you call it when an object speeds up, slows down, or changes direction? If you said a "change in velocity," you would be correct. What is another word for a change in velocity in a particular period of time? Acceleration. Why would the spaceship need to burn fuel to accelerate? Newton's first law tells us that in the absence of an unbalanced force, an object in motion will continue in uniform motion. What is the effect of an unbalanced force on the same moving object? An acceleration.

Newton's second law of motion, the law of acceleration, states that the acceleration an object experiences is directly proportional to the unbalanced force acting on it, and inversely proportional to its own mass. Mathematically, this statement could be expressed with the following formula:

$$\text{acceleration of an object} = \frac{\text{net force on the object}}{\text{mass of the object}} \quad \text{or,} \quad a = \frac{F_{net}}{m}.$$

Once again, we can think of many real-life examples that support this law. If you push hard on an object, it moves away from you faster than if you push gently on it. That is because the acceleration that the object experiences is directly proportional (as one goes up, the other goes up) to the force you exert on it. If you use the same amount of force to push a light object as a heavy object, the light object will move away from you with a greater velocity. This is because the acceleration that an object experiences is inversely proportional (as one goes up the other goes down) to the mass of the object. Both acceleration and force are vector quantities, requiring numbers, units, and direction to describe them completely. It should also be noted that the acceleration that an object experiences is in the same direction as the net force acting on it.

Newton's second law is usually shown with the symbol for force isolated on one side of the equal sign, as in $F_{net} = ma$ or simply $F = ma$, where the F is assumed to mean "net force." This seemingly simple formula is very important to your study of physics. Tattooing it onto your body might be a little extreme, but you should become so familiar with it that you come to know it like "the back of your own hand." It is that important!

Newton's Second Law: The Law of Acceleration

The acceleration an object experiences is directly proportional to the net unbalanced force exerted upon it and inversely proportional to its mass.

$$F = ma$$

In Lesson 2–1 we defined a force as a push or a pull. Newton's second law shows us that we can also define a force as the product of mass and acceleration. When we multiply the units of mass (kg) by the units of acceleration (m/s^2) we get $kg \cdot m/s^2$. Rather than write all of these units every time we describe a force, a derived unit called the newton (N) has been created. It is also important to keep in mind that force is a vector unit, so a direction must be included in a complete description. In many problems, however, no direction will be mentioned. In these cases you are only working with the magnitude (numbers and units) of the force, not direction.

Many of the problems that you encounter in your physics class can be solved with, or at least partially solved with, Newton's second law. In fact, there is a good chance that there is no other formula that you will use more in physics this year than this one! If your teacher calls on you in class and asks how you should go about solving a motion problem that you don't know how to start, try saying, "Can we use Newton's second law of motion?" Chances are, you probably can! Let's go over some examples of problems that can be solved with this formula.

Example 1

A crate with a mass of 50.0 kg experiences an acceleration with a magnitude of 3.50 m/s². Find the magnitude of the net force on the crate.

Given: m = 50.0 kg   a = 3.50 m/s²

Find: F

Solution: $F = ma = (50.0\ \text{kg})(3.50\ \text{m/s}^2) = 175\ \text{kg} \times \text{m/s}^2 = \mathbf{175\ N}$

That should be easy enough! Remember, as with all of the formulas that you use this year, different variables can be the unknown in a problem. Let's try an example where acceleration is the unknown.

Example 2

A boy pushes a 0.500 kg book across a table by exerting a net force of 2.00 N on it. Calculate the magnitude of the acceleration that the book will experience.

Given: m = 0.500 kg   F = 2.00 N

Find: a

Isolate:

Starting with: F = ma

Divide both sides by m: $\frac{F}{m} = \frac{\cancel{m}a}{\cancel{m}}$

We get: $a = \frac{F}{m}$

Solution: $a = \frac{F}{m} = \frac{2.00\ \text{N}}{0.500\ \text{kg}} = \mathbf{4.00\ m/s^2}$

Weight

Now that we have gone over the basic type of problem, let's go over a couple of special cases. First, you need to understand that what we call "weight" is a specific example of a force. **Weight** is a measure of the force of gravity between two objects, one of which is usually Earth. If we know the mass of an object, we can determine its weight in newtons by multiplying its mass by its acceleration. Which acceleration do we use? The acceleration due to gravity ($g = -9.81\ m/s^2$), provided that we want the weight of the object on Earth.

Formula for Calculating the Weight of an Object on Earth

$$F_w = mg$$

Understand that this is not a new formula. Rather, it represents a specific case for finding a specific force (force of weight) using a specific acceleration (the acceleration due to gravity). You will be required to use this formula many times this year, so make sure you know when to use it!

Example 3

Find the weight of a person (on Earth) in newtons if they have a mass of 55.0 kg.

Given: $m = 55.0\ kg$ $\quad g = -9.81\ m/s^2$

Find: F_w

Solution: $F_w = mg = (55.0\ kg)(-9.81\ m/s^2) = -539.55\ kg \cdot m/s^2$

$= \mathbf{-5.40 \times 10^2\ N}$

Do you see why we expressed our answer in scientific notation? I wanted to round to three significant digits, which would force me to round to 540. Using scientific notation allowed me to show three significant digits. Why is the weight in our answer negative? Does that mean that the object is very light? No, the negative sign simply indicates that the direction is downward. Your instructor may have a preference about including the sign in the answer. You will often see weights listed without signs because we are often interested only in the magnitude of the weight, and the direction is understood.

You should keep in mind that the weight of an object is proportional to its mass, but mass and weight are not the same thing. Your weight will change due to location, but your mass won't. Sometimes a question is designed to test whether or not a student distinguishes between mass and weight. Try Example 4 by yourself, but be careful with it. Some students call this a trick problem, but instructors use this type of problem to test whether or not their students are paying attention to units.

Example 4

A 40.0 N brick experiences a net force 5.00 N parallel to its surface. Find the magnitude of the acceleration that the block will experience.

When the problem refers to a "40.0 N brick" it means that the weight of the brick is 40.0 N. We know that this can't be the mass of the brick, because the units of mass are kilograms, not newtons. If we want to use Newton's second law to find the acceleration of the brick, we must first use the special case of Newton's second law to find the mass of the object.

Convert: $m = \frac{F_w}{g} = \frac{40.0\ \text{N}}{9.81\ \text{m/s}^2} = \mathbf{4.08\ kg}$

Next, use the mass of the brick to find the acceleration.

Given: $m = 4.08$ kg $\qquad F_{net} = 5.00$ N

Find: a

Solution: $a = \frac{F_{net}}{m} = \frac{5.00\ \text{N}}{4.08\ \text{kg}} = \mathbf{1.23\ m/s^2}$

Now, try the following review problems.

Lesson 2–3 Review

1. ______________ is the force of attraction between two objects due to gravity, where one object is often Earth.
2. What would be the mass of an object that has a weight of 295 N near the surface of Earth?
3. Find the magnitude of the acceleration experienced by a 55.0 kg object with an unbalanced force of 125 N acting on it.

Lesson 2-4: Newton's Third Law of Motion

You have probably heard and may have even used the expression "for every action, there is an equal and opposite reaction." This expression is actually a rephrasing of Newton's third law of motion, the law of action-reaction. There is something that bothers me about this expression, because I believe it leads to a certain amount of confusion on the part of students. When a baseball bat hits a baseball, the bat and the ball don't seem to react the same way. Better yet, when a car hits a deer in the road, the car doesn't seem to react in the same way as the deer. The car may get damaged and slow down a small amount, but the deer probably gets killed, and may even be thrown some number of feet. How are these "equal and opposite" reactions? Does Newton's third law not apply in this case?

Newton's third law actually states that if an object exerts a force on a second object, the second object exerts an equal and opposite force on the first object. It is actually the forces exerted between two objects that are equal and opposite, not what we might think of as the "reactions" of the objects.

Newton's Third Law: The Law of Action-Reaction

When an object exerts a force on a second object, the second object exerts an equal and opposite force on the first object.

Newton's third law can be seen in countless real-life situations. For example, if you punch your brother in the face (and I am certainly not recommending it), in addition to getting grounded by your parents, you will probably end up with a hurt fist. Why does it hurt your hand when you punch something or someone? Newton's third law. You can't touch without being touched. If your fist exerts a force on his face, his face will exert an equal and opposite force on your fist. This is an example of what we will refer to an **action-reaction pair** of forces.

In our example with a baseball and a bat, the bat exerts a force on the ball and the ball exerts an equal and opposite force on the bat. For example, if the bat exerts a 20.0 N force due east on the ball, the ball exerts a 20.0 N force due west on the bat. The two objects don't "react" to these respective equal forces in the same way, because each object has its own mass. The mass of the bat is much greater, so it doesn't experience as great an acceleration as the ball, as explained by Newton's second law, $F = ma$. Let's look at this example more closely.

Example 1

A baseball bat with a mass of 5.00 kg exerts a force of 20.0 N on a 0.250 kg baseball. Find the resultant acceleration of each object.

This question requires we use both Newton's second and third laws to solve it. The third law tells us that if the bat exerts a 20.0 N force on the ball, the ball exerts a 20.0 N force on the bat. To show that the forces are in opposite directions, we will call one positive and one negative. It is arbitrary which force we call negative, but I will choose to call the "reaction" force, that is the force exerted by the ball on the bat, negative.

Given: $m_{bat} = 5.00$ kg $\quad m_{ball} = 0.250$ kg $\quad F_{bat\ on\ ball} = 20.0$ N
$F_{ball\ on\ bat} = -20.0$ N

Find: a_{bat} a_{ball}

Solution: $$a_{bat} = \frac{F_{ball\ on\ bat}}{m_{bat}} = \frac{-20.0\text{ N}}{5.00\text{ kg}} = \mathbf{-4.00\ m/s^2}$$

$$a_{ball} = \frac{F_{bat\ on\ ball}}{m_{ball}} = \frac{20.0\text{ N}}{0.250\text{ kg}} = \mathbf{80.0\ m/s^2}$$

As you can see, both objects experience equal and opposite forces, but they react to these forces in very different ways. The ball experiences a much greater acceleration because it has much less mass. The bat is more massive, so a smaller acceleration results. The fact that the accelerations are in different directions makes sense, because each object will experience an acceleration in the direction of the net force on it. The ball will actually change directions and head in the direction of the force applied *by* the bat, while the bat only slows down in the direction of the force applied *on* the bat.

When a book is placed on a table, it will exert a downward force on the table that is equal to its own weight. The table must exert an equal and opposite force on the book, according to Newton's third law. If the table exerted a reactionary force that was less than the weight of the book, the book would break the table and fall towards the ground. If the table exerted a reactionary force that was greater than the weight of the book, the book would actually be pushed upwards, off of the table.

Example 2

A book with a mass of 1.50 kg rests on a table. Find the weight of the book (F_w) and the force being exerted by the table on the book (F_N).

Given: m = 1.50 kg $\qquad$ g = −9.81 m/s²

Find: F_w and F_N

Solution: The weight of the book:

$F_w = mg = (1.50\ \text{kg})(-9.81\ \text{m/s}^2) = \mathbf{-14.7\ N}$

The negative sign in the answer simply indicates that the force is directed downwards. This force of weight is equal to the force that the book will exert on the table. The table must be exerting an equal and opposite force, so

$$F_N = F_w = \mathbf{14.7\ N}$$

where the positive value indicates that the force is directed upwards.

It is important to point out that, while the force exerted by the table on the book in Example 2 was equal and opposite to the book's weight, these forces do not represent action-reaction force pairs. The force pairs are the force exerted by the book on the table and the force exerted by the table on the book. What is the force that is the equal and opposite force to the force of weight?

One of the most profound aspects of Newton's third law comes from the understanding that this law holds true for all types of forces, including forces at a distance. This means that when Earth exerts a force on you, due to gravity, you exert an equal and opposite force on Earth! If you were to jump off your bed, you would fall to the floor because Earth is pulling you down with a force equal to your weight in newtons. What you might not realize is that you are also pulling the Earth up with an equal and opposite force! Newton's laws tell us that you have the power to move planets!

"Impossible!" you say? Suppose I tell you that the mass of a boy is 60.0 kg. You can use this information and the acceleration due to gravity (−9.81 m/s²) to find the boy's weight, as described in our last lesson. This weight is the force exerted on the boy by Earth, due to gravity.

The boy's weight: $F_w = m_b g = (60.0\text{ kg})(-9.81\text{ m/s}^2) = \mathbf{-589\text{ N}}$

The negative sign in this case simply tells us that the force is being directed downwards, or towards the center of Earth. Newton's third law tells us that if Earth is exerting a force on the boy, the boy must be exerting an equal and opposite force on Earth. So, the boy is also exerting a **589 N** force on Earth. For the sake of simplicity, let's pretend that this represents an unbalanced, or net, force. Now, if I tell you that the mass of Earth (m_e) is 5.98×10^{24} kg, can you determine the acceleration of Earth as a result of the reaction force to the boy's weight? Let's try the following calculation together.

Example 3

A 589 N boy jumps off his bed. As he falls to the floor, he accelerates downwards at a rate of -9.81 m/s^2 due to gravity. Earth has a mass (m_e) of approximately 5.98×10^{24} kg. Use this information to find the rate of the acceleration of Earth upwards towards the boy (a_e).

Given: $F_e = 589\text{ N}$ $m_e = 5.98 \times 10^{24}\text{ kg}$

Find: a_e

Solution:

$$a_e = \frac{F_e}{m_e} = \frac{589\text{ N}}{5.98 \times 10^{24}\text{ kg}} = \mathbf{9.85 \times 10^{-23}\ m/s^2}$$

$$= \mathbf{0.000\,000\,000\,000\,000\,000\,000\ \ 009\,85\ m/s^2}$$

Do you see how insignificant the acceleration of Earth would be in this situation? Combine that with the fact that the boy is in the air for a fraction of a second, and you can see why the boy's effect on the planet Earth is not noticeable. The force really is there, but a Ping-Pong ball has a better chance of knocking down a house than the boy does of causing problems for Earth.

Let's try some practice questions before moving on to the next lesson.

Lesson 2–4 Review

1. If a rock with a mass of 2.45 kg rests on a horizontal driveway, what is the magnitude of the force that the driveway is exerting on the rock?

2. According to Newton's third law, if a car hits a garbage can, will they both experience the same acceleration?
3. Use Newton's third law to explain what happens when someone breaks through a thin layer of ice on a pond.

Lesson 2–5: Motion Along a Horizontal Surface

In this lesson we get to cover some of my favorite problems. I always enjoy teaching this section because, at this point, the problems start getting sophisticated enough to make them pretty interesting. To solve these problems involving motion on a horizontal surface, students need to make use of a number of skills that they practiced earlier. It becomes necessary to move away from the mechanical type of problem-solving that may feel "safe," to the more challenging type of problem-solving, which requires logic and understanding.

Invariably, when students get to this lesson they want a formula that they can use in all situations. Unfortunately, while most of these problems all come down to Newton's second law, there is no "master formula" that can be used for every situation. You need to learn to analyze each particular situation and choose the appropriate formulas. It is analogous to a mechanic or carpenter choosing the appropriate tool for a given situation. In this way, each new formula that you learn is like a new tool that you can add to your mental toolbox, for use when you need it.

Coefficient of Friction

Let's add another tool to that box right now. Have you ever been walking on a driveway and you suddenly step on a patch of ice and slip? What is the difference between the two surfaces? There is less friction between your foot and the ice. Your foot slips more freely, and you fall as a result.

But what is friction? We did an example of a problem with friction in Lesson 2–1, but we didn't really go into much detail. **Friction** is a force between two objects that are in contact with each other, exerted parallel to the surface of contact. When the two objects are not moving relative to each other, this force is called **static friction**. When one object is moving relative to the other, this force is called **kinetic friction**. You may recall a situation where you have noticed the difference between these two kinds of friction. Have you even needed to push a heavy object, say a desk or dresser, across a floor? Did you notice that it is hard to start the object moving, but once the object is in motion, it takes less effort to keep the object moving? This is because the force of kinetic friction is always less than the force of static friction, so the force you needed to overcome to get the object moving is less than the force that you need to overcome to keep it moving.

What determines the magnitude of the friction between two objects? The first thing, as you might imagine from our examples involving ice, is the nature of the materials in contact. The second has to do with the normal (perpendicular) force between the objects, which you might think of as how hard the objects are pressed together. If you touch your hands lightly together and rub them back and forth, they slide more freely than if you press them together harder. In the same way, there is more friction between a heavy desk and a wood floor than there would be if you emptied the drawers of the desk and made it lighter.

When you divide the force of static friction by the normal force between two stationary objects that are in contact, you get the **coefficient of static friction**, as shown here:

$$\text{Coefficient of static friction}\,(\mu_s) = \frac{\text{force of static friction}\,(F_f)}{\text{normal force}\,(F_N)}.$$

When you divide the force of kinetic friction by the normal force between objects that are in contact and motion relative to each other, you get the **coefficient of kinetic friction**, as shown here:

$$\text{Coefficient of kinetic friction}\,(\mu_k) = \frac{\text{force of kinetic friction}\,(F_f)}{\text{normal force}\,(F_N)}.$$

The greater the force of friction between two objects, the higher the coefficient of friction will be.

Let's try a couple examples that deal with coefficient of friction.

Example 1

A desk with a mass of 75.0 kg sits on a floor, and the coefficient of static friction between the two surfaces is 0.40. How much force must someone apply in a direction horizontal to the surface of the floor to get the desk moving?

Can you see how to solve this problem? We will use the mass of the desk and the acceleration due to gravity to find the weight of the desk.

The magnitude of this weight will be equal to the normal force. We will then use the normal force and the coefficient of static friction to determine the maximum value of the force of static friction. The force required to move the desk will be greater than this maximum value for the force of static friction.

F_N F_f F_a 75.0 kg $\mu = 0.40$ F_w

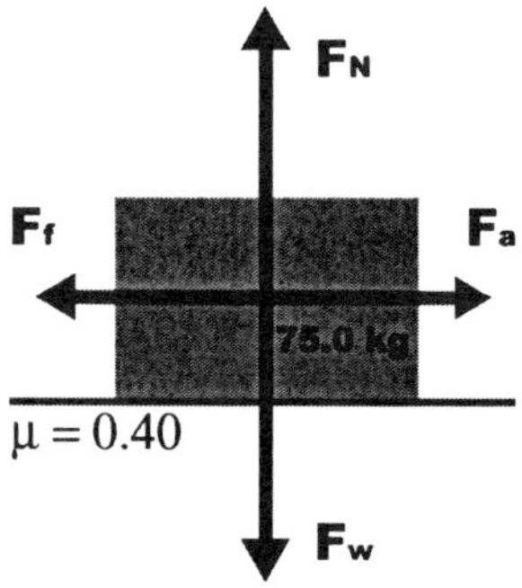

Figure 2.6

Given: $m = 75.0$ kg $\quad g = 9.81$ m/s^2
$\mu_s = 0.40$

Find: F_N and then F_f

Normal force $= F_N = -F_w = -mg = -(75.0 \text{ kg})(-9.81 \text{ m/s}^2) = \mathbf{736\ N}$

Static friction $= F_f = \mu_s F_N = (0.40)(736 \text{ N}) = \mathbf{290\ N}$

The desk will begin to slide if a force that is greater than this maximum value of the force of static friction is applied, so our answer is **greater than 290 N**.

Now let's try a slightly more complicated problem.

Example 2

A boy exerts a force of 12.0 N on a 20.0 N box to slide it across a table where the coefficient of kinetic friction between the surfaces is 0.34. Find the acceleration on the box.

Given: $F_w = 20.0$ N $\quad g = 9.81$ m/s^2 $\quad F_a = 12.0$ N
$\mu_k = 0.34$

Find: a

Can you come up with a plan to solve this problem? They ask for the acceleration, so we can use Newton's second law ($F_{net} = ma$) for that. We will need to find the mass of the box, but that will be as easy as dividing the weight of the box by the acceleration due to gravity, g. The normal force is easy as well, because in this case it will be equal to the magnitude of the weight of the box. To find the net force (F_{net}) we will first need to find the force of friction (F_f) using the coefficient of kinetic friction and the normal force. We give the friction a negative sign because it opposes the motion of the box. If we gave the force of friction the same sign as the applied force, the friction would actually help the boy accelerate the box!

Solution:

Mass of the box = $m = \frac{F_w}{g} = \frac{20.0\text{ N}}{9.81\text{ m/s}^2} = \mathbf{2.04\text{ kg}}$

Force of kinetic friction = $F_f = \mu_k F_N = \mu_k F_w$
$= (0.34)(20.0\text{ N}) = \mathbf{6.8\text{ N}}$

Net force = $F_{net} = F_a + F_f = 12.0\text{ N} + (-6.8\text{ N}) = \mathbf{5.2\text{ N}}$

$$a = \frac{F_{net}}{m} = \frac{5.2\text{ N}}{2.04\text{ kg}} = \mathbf{2.5\text{ m/s}^2}$$

The fact that the acceleration has the same sign and therefore the same direction as the applied force means that the box will speed up. If the applied force were less than the force of kinetic friction, the box would slow down.

Now that we have gone over the basics, let's try a more difficult problem. Let's imagine a situation where a girl is dragging a sled over the ground by exerting a force on a rope that makes an angle above the horizontal. In such a situation, not all of the force that the girl is exerting is parallel to the ground, opposite to the friction. We must resolve the original applied force into its *x* and *y* components. The *x* component of the force will be working against the force of friction, while the *y* component of the applied force will work against the weight of the sled, decreasing the normal force, which will, in turn, decrease the force of friction.

Example 3

A girl is pulling a sled with a mass of 16.0 kg across a level patch of grass by exerting a force of 22.0 N along a rope that forms an angle of 33.0° above the ground. If the coefficient of kinetic friction between the sled and the grass is 0.25, find the acceleration of the sled.

Given: $m = 16.0\text{ kg}$ $F_a = 22.0$ N at 33.0° above the horizontal
$\mu_k = 0.25$

Find: a

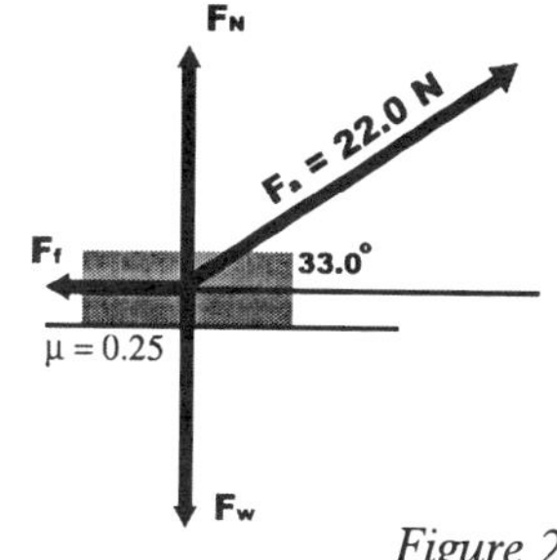

Figure 2.7

As I mentioned in the introduction to this lesson, there is a lot to some of these problems. You can't just look up a formula and plug in the given information. You need to slow down, think, and formulate a plan for solving them. This problem asks us to find the acceleration of the sled, so we know that we will be using Newton's second law. That means that we will need to find the net force on the sled. In our previous example, the force of weight and the normal force were equal and opposite, leaving us with a net force of zero in the vertical, so we only considered the force parallel to the ground. Finding the net force in this problem is a bit more complicated, because now we must consider an extra vertical force, the *y* component of the applied force. As the girl pulls on the rope, she is taking on some of the weight of the sled, decreasing the force it will exert on the ground, and the normal force. We will still end up with a net force of zero perpendicular to the ground, but we must take into account how the vertical component of the applied force affects the force of kinetic friction.

Solution:

Let's begin by finding the *x* and *y* components of the applied force, using the trigonometry functions with which you should now be familiar.

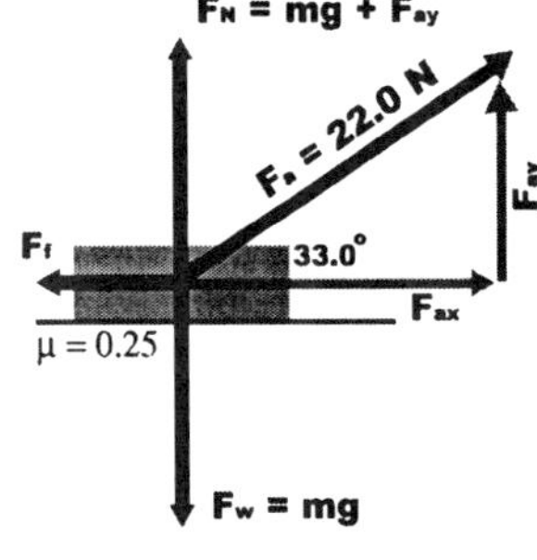

Figure 2.8

Force applied in the $x = F_{ax} = F_a \cos\theta$
$= (22.0\text{ N})(\cos 33.0^\circ) =$ **18.5 N**

Force applied in the $y = F_{ay} = F_a \sin\theta$ $= (22.0\text{ N})(\sin 33.0^\circ) =$ **12.0 N**

Now let's find the normal force. Remember, the force of weight is directed downwards, and the vertical component of the applied force is directed upwards. When we add these forces let's make the force applied in the *y* negative, so they work against each other.

Normal force $= F_N = -F_w + F_{ay} = mg + F_{ay}$
$= -(16.0\text{ kg})(-9.81\text{ m/s}^2) + (-12.0\text{ N}) =$ **145 N**

We can use the normal force and the coefficient of kinetic friction to find the force of kinetic friction.

Force of kinetic friction = $F_f = \mu_k F_N$ = (0.25)(145 N) = **36.3 N**

We will add the force applied in the *x* to the force of friction to find the net force acting on the sled. Of course, because the force of friction is working against the applied force, we will make it negative.

Net force = $F_{net} = F_{ax} + F_f$ = 18.5 N + (–36.3 N) = **–17.8 N**

Remember, the negative value of our answer tells us that the force of friction is greater than the force the girl is applying in the horizontal, and in the opposite direction. This means that the sled is going to be slowing down. We find the acceleration here:

$$\text{Acceleration} = a = \frac{F_{net}}{m} = \frac{-17.8\text{ N}}{16.0\text{ kg}} = \mathbf{-1.11\ m/s^2}$$

Will the sled stop? That depends on whether or not the girl can apply a greater force, or if she moves off the grass onto another surface. What would happen if she changes the angle of the rope instead of changing the force she exerts? Decreasing the angle would increase the component of the force along the horizontal, but it would also decrease the force exerted in the vertical. Decreasing the force exerted in the vertical will actually increase the normal force and the force of friction! Would decreasing the angle between the rope and the ground make it easier or harder for the girl to pull the sled over the grass? Let's see what happens if she changes the angle between the rope and the horizontal to 10.0°. This would be a good opportunity for you to try the problem on paper, and check your answer against mine.

Example 4

A girl is pulling a sled with a mass of 16.0 kg across a level patch of grass by exerting a force of 22.0 N along a rope that forms an angle of 10.0° above the ground. If the coefficient of kinetic friction between the sled and the grass is 0.25, find the acceleration of the sled.

Given: m = 16.0 kg $\quad F_a$ = 22.0 N at 10.0° above the horizontal
μ_k = 0.25

Find: a

Force applied in the *x* = $F_{ax} = F_a \cos\theta$
= (22.0 N)(cos 10.0°) = **21.7 N**

Force applied in the $y = F_{ay} = F_a \sin\theta$
$= (22.0\ \text{N})(\sin 10.0^\circ) = \mathbf{3.82\ N}$

Normal force $= F_N = -F_w + F_{ay} = mg + F_{ay}$
$= -(16.0\ \text{kg})(-9.81\ \text{m/s}^2) + (-3.82\ \text{N}) = \mathbf{153\ N}$

Force of kinetic friction $= F_f = \mu_k F_N = (0.25)(153\ \text{N}) = \mathbf{38.3\ N}$

Net force $= F_{net} = F_{ax} + F_f = 21.7\ \text{N} + (-38.3\ \text{N}) = \mathbf{-16.6\ N}$

Acceleration $= a = \dfrac{F_{net}}{m} = \dfrac{-16.6\ \text{N}}{16.0\ \text{kg}} = \mathbf{-1.04\ m/s^2}$

The sled is still slowing down, but at a lesser rate. So decreasing the angle between the rope and the ground would benefit the girl.

You may want to practice this problem a few times, changing the angle of the rope or the applied force each time. Keep in mind that the acceleration will not always be the unknown. Sometimes the coefficient of friction is the unknown and the acceleration is given. You need to learn to adapt your train of thought to the given situation.

When you feel ready, go on to the review problems.

Lesson 2–5 Review

1. ________________ is the force of friction between two objects that are not in motion relative to each other.

2. ________________ is the ratio of the force of static or kinetic friction to the normal force.

Lesson 2–6: Motion on an Inclined Surface

In this lesson we study motion on an "inclined plane," which is a fancy term for a ramp. There are many real-life applications for this type of problem. Movers will sometimes use a ramp to avoid having to lift a heavy object straight up. Motion on an inclined surface can be seen at the park, where children slide down slides. Driving a car up a hill represents motion on an incline. Skiing and sleigh riding often take place on an incline as well.

Here's a little experiment that you can try right now. Find a board or some other sturdy object that can serve as a ramp. Place a book or some other object on the board when it is lying flat.

Slowly lift one end of the board and see how steep you need to make the incline before the book starts to slide slowly down it. Now, ask yourself the important question. What makes the book slide down the ramp? The book was at rest, and then it started to move.

Newton's second law tells us that an unbalanced force is required to set the book in motion. Where did the force come from? If you said "gravity," you are correct. But the force of gravity was always there, even when the board was lying horizontally. The force of gravity between Earth and the book didn't change as you changed the angle of incline of the board. So, what did change?

It will help to draw a free-body diagram for the book when it was lying on a horizontal surface, and another for when it was lying on an inclined surface.

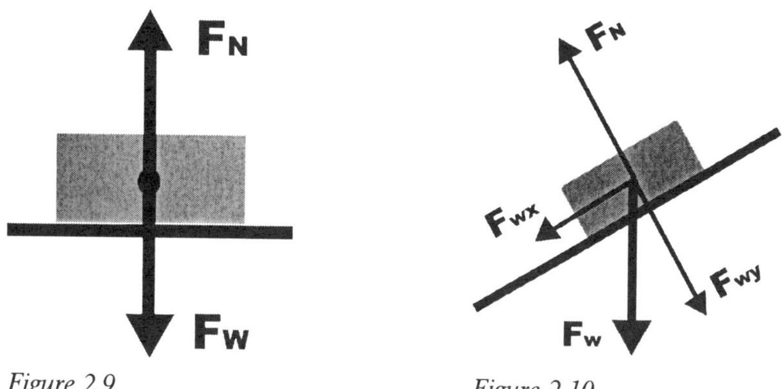

Figure 2.9

Figure 2.10

What you see is that when the book was on a horizontal surface, the normal force exerted by the board on the book was equal to the weight of the book and in the opposite direction. As you lifted one end of the board, the weight of the book is still directed straight down, but the normal force is still perpendicular to the surface of the board. If you break the force of weight of the book into two components, the normal force is still equal and opposite to the component that is perpendicular to the surface, but another component of the weight is shown parallel to the surface of the incline. Even when the book was on a slight incline, it probably didn't move, because the parallel component of the weight is balanced by the force of static friction up to a point. When the incline gets so great that the parallel component of the book's weight is greater than the maximum value for the force of static friction, the book begins to slide down the incline.

You can also use this setup to test the fact that the value for kinetic friction is less than the value for static friction. Lay the board flat again, put the book on it, and then raise it to an angle that is just less than the angle that caused it to slide last time. Tap the book, and it should slide down the board, accelerating along the way. The fact that it didn't slide before you tapped it shows that the force of static friction was not overcome until you tapped it. Once in motion, it accelerated because the parallel component of the weight was greater than the value of kinetic friction.

As you can imagine, the materials in contact will influence the angle that the board must reach before the object begins to slide. A different object, such as a glass, might have started to slide when the board was at a lesser angle. This is because there is less friction between some items. The coefficient of friction must be taken into account when dealing with most of these problems.

One way to approach inclined plane problems is to realign our x and y coordinate system so that the x-axis lines up parallel to the incline and the y-axis is perpendicular to the incline.

Let us ignore friction for a moment and begin with the most basic of inclined plane problems. For this problem, I am going to ask you to ignore the friction between the inclined plane and the object we place on it. Imagine how slippery a steep hill covered by a sheet of ice can be, and then take it further.

Example 1

A block with a mass of 5.00 kg is placed on a frictionless inclined plane at an angle of 48.0° above the horizontal. Draw a diagram for the problem and find the acceleration of the block.

Given: m = 5.00 kg $\theta = 48.0°$
g = 9.81 m/s²

Find: a

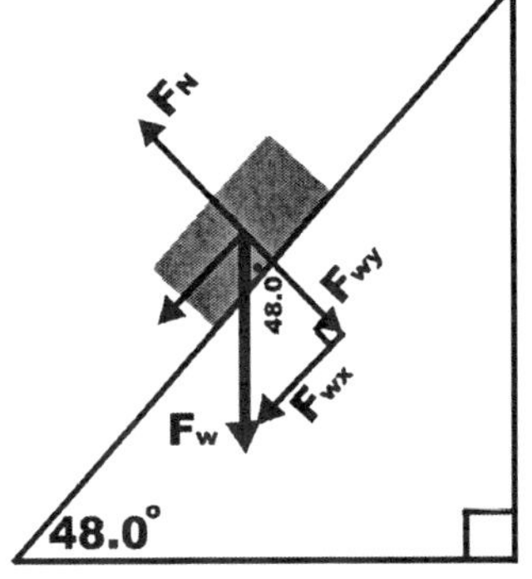

Figure 2.11

Study Figure 2.11. It is likely that you will need to produce similar diagrams, and it is important that you understand how they are constructed.

In our diagram the larger triangle represents the ramp, or inclined plane. The smaller triangle is formed from the vectors representing the force of weight of the block and the component vectors of the weight. The weight of the block (F_w) is directed straight down towards the center of Earth, ignoring the angle of the incline. We then find the components of the weight that we are interested in. The *y* component (F_{wy}) is the component of the weight that is perpendicular to the surface of the incline, and the *x* component (F_{wx}) is the weight that is parallel to the surface of the incline. The larger and smaller triangles are similar triangles, meaning they have the same angles. Therefore, the angle between the vector representing the weight and the vector representing the perpendicular component of the weight is also 48.0°.

The *y* component of the weight (F_{wy}) will not cause the block to slide down the ramp, as it is equal and opposite to the force that the ramp exerts on the block. Therefore, the net force perpendicular to the surface of the ramp is zero. The *x* component of the weight (F_{wx}), on the other hand, will cause the block to slide down the ramp, provided it is greater than the force of friction. Let's calculate the weight of the block using Newton's second law, and the component of the weight that is parallel to the ramp.

$$\text{Weight} = F_w = mg = (5.00\text{ kg})(9.81\text{ m/s}^2) = \mathbf{49.1\ N}$$

$$\text{Parallel Force} = F_{wx} = F_w \sin\theta = (49.1\text{ N})(\sin 48.0^\circ) = \mathbf{36.5\ N}$$

Can you see why we multiplied the weight by the sine 48.0°? The weight forms the hypotenuse of our inner triangle, while the parallel component of the weight appears as the opposite side to the 48.0°. So, we use the relationship between the hypotenuse and the opposite side, given by

$$\sin\theta = \frac{\text{opposite}}{\text{hypotenuse}}.$$

Of course, we could have done the calculation in one step, and many students prefer to remember the formula for the *x* component of the weight as $F_{wx} = mg\sin\theta$.

Now we have everything we need to find the acceleration of the block down the incline. Because there is no friction, the only unbalanced force acting on the block is the parallel component of its weight. Therefore, the net force is the same as the parallel component of the weight. We can use Newton's second law to find the acceleration.

$$\text{acceleration (a)} = \frac{F_{net}}{m} = \frac{F_p}{m} = \frac{36.5\text{ N}}{5.00\text{ kg}} = \mathbf{7.30\ m/s^2}$$

You may have noticed that we multiplied g by the mass of the block in our first calculation, and then we divided the F_p by the mass in our last calculation. These steps actually cancel each other out! I went through each of these steps to explain to you logically what we are doing, but now I will show you the easier way to do this calculation.

$$\text{acceleration (a)} = \frac{F_{net}}{m} = \frac{F_{wx}}{m} = \frac{F_w \sin\theta}{m} = \frac{\cancel{m} g \sin\theta}{\cancel{m}} = \mathbf{g\sin\theta}$$

Using this simplified formula to solve Example 1 we would get:

$$\text{acceleration (a)} = g\sin\theta = (9.81\ \text{m/s}^2)(\sin 48.0^\circ) = \mathbf{7.29\ m/s^2}.$$

So, we can get the same answer (with a slight difference due to rounding) with far less work! Furthermore, we don't need to know the mass of the object to do the calculation. The key to remember is that the acceleration of an object sliding down a frictionless ramp is simply gsin θ.

Example 2

A crate slides down a frictionless ramp with an angle of 33.5° above the horizontal. Find the acceleration of the crate.

Given: $g = 9.81\ \text{m/s}^2$ $\theta = 33.5^\circ$

Find: a

Solution: $a = g\sin\theta = (9.81\ \text{m/s}^2)(\sin 33.5^\circ) = \mathbf{5.41\ m/s^2}$

Perhaps you are thinking, *How often am I going to encounter a ramp with a frictionless surface in real life?* Your point is well taken! Let's try another example where we have a coefficient of friction between the block and the surface of the ramp.

Example 3

A crate with a mass of 35.0 kg is sliding down an inclined plane with an angle of 43.0° above the horizontal. If the crate is accelerating at a rate of 3.45 m/s², find the coefficient of kinetic friction between the ramp and crate.

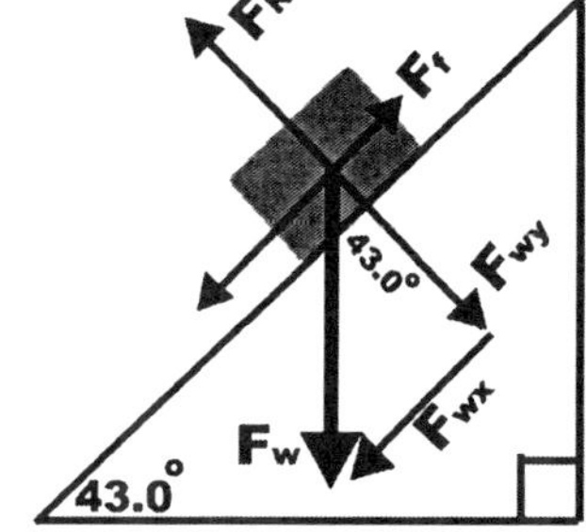

Figure 2.12

Given: $m = 35.0$ kg $\quad g = 9.81$ m/s^2 $\quad a = 3.45$ m/s^2
$\theta = 43.0°$

Find: μ_k

As a beginner, it is hard to know where to start these problems. You consider the information that you have been given and the formulas that you know, and you may not be sure how to get to what you want to find. There are actually several different ways to proceed, but I will show you how to work backwards from what you need to what you have. I will then show you how to put all of the formulas together to solve the problem in one step.

Our unknown is the coefficient of kinetic friction. From Lesson 2–5, we have the formula for determining the coefficient of kinetic friction:

$$\text{Coefficient of kinetic friction } (\mu_k) = \frac{\text{force of kinetic friction } (F_f)}{\text{normal force } (F_N)},$$

$$\text{or } \mu_k = \frac{F_f}{F_N}.$$

We can see from Figure 2.12 that the *y* component of the weight (F_{wy}) represents the adjacent side of our inner triangle, and this will be equal to the normal force (F_N). The hypotenuse of our inner triangle is the weight (F_w) of the crate, which we find with Newton's second law, $F_w = mg$. We can use the relationship between the adjacent side (F_{wy}) and hypotenuse (F_w) of our inner triangle, shown by $\cos\theta = \frac{\text{adjacent side}}{\text{hypotenuse}} = \frac{F_{Wy}}{F_w}$ to find the normal force. That means that the normal force is equal to

$F_N = -F_{wy} = -F_w \cos\theta = -mg\cos\theta = -(35.0 \text{ kg})(-9.81 \text{ m/s}^2)(\cos 43.0°)$
$= \mathbf{251\ N.}$

That takes care of the denominator in our working formula.

$$\mu_k = \frac{F_f}{F_N} = \frac{F_f}{251\text{ N}}$$

Now, all we need to do is figure out the force of kinetic friction between the crate and the ramp. From our diagram (Figure 2.12), we can see that there are two forces acting on the crate that are parallel to the surface of the ramp. The parallel component of the force of weight (F_{wx}) is causing the crate to slide down the ramp, while the force of kinetic friction (F_f) is working against this force.

As we learned earlier in this lesson, the parallel component (F_{wx}) of the weight of an object on an inclined plane can be found with the formula $F_{wx} = mg\sin\theta$. We have everything we need to find the F_{wx}:

$$F_{wx} = mg\sin\theta = (35.0\text{ kg})(9.81\text{ m/s}^2)(\sin 43.0^\circ) = \mathbf{234\ N}.$$

We can use Newton's second law, the acceleration of the crate, and the mass of the crate to find the net force (F_{net}) acting on it.

$$F_{net} = ma = (35.0\text{ kg})(3.45\text{ m/s}^2) = \mathbf{121\ N}$$

The net force (F_{net}) acting on this crate is going to be the sum of the two forces, $F_{net} = F_{wx} + F_f$. If we isolate the force of friction in this equation, we get $F_f = F_{net} - F_{wx}$. Solving for the force of friction, we find:

$$F_f = F_{net} - F_{wx} = 121\text{ N} - 234\text{ N} = \mathbf{-113\ N}.$$

This gives us the numerator for our formula for the coefficient of kinetic friction, so we can get our final answer. We will use only the magnitude of the force of friction:

$$\mu_k = \frac{F_f}{F_N} = \frac{113\ \cancel{\text{N}}}{251\ \cancel{\text{N}}} = \mathbf{0.450}\ .$$

That was a lot of work, but I hope I didn't lose you. With some practice, you should come to enjoy the puzzle-solving aspect of these problems. If you prefer to come up with one "master formula" for the problem and solve it in one calculation, it might look like the following:

$$\mu_k = \frac{F_f}{F_N} = \frac{F_{net} - F_p}{F_w\ \cos\theta} = \frac{\cancel{m}a - (\cancel{m}g\sin\theta)}{\cancel{m}g\cos\theta} = \frac{a - (g\sin\theta)}{g\cos\theta}$$

$$= \frac{3.45\ \cancel{\text{m}}/\cancel{\text{s}}^2 - (9.81\ \cancel{\text{m}}/\cancel{\text{s}}^2)(\sin 43.0^\circ)}{(9.81\ \cancel{\text{m}}/\cancel{\text{s}}^2)(\cos 43.0^\circ)} = \mathbf{0.451}$$

Please note, the slight difference in the two answers is due to rounding, and is not significant.

Chapter 2 Examination

Part I—Matching

Match the following terms to the definitions that follow.

a. force d. contact force g. field force

b. net force e. weight h. inertia

c. friction f. static friction i. kinetic friction

_____1. A force that exists between two objects that are touching, which resists their motion relative to each other.

_____2. The force of attraction between two objects due to gravity.

_____3. The sum of all of the forces acting on an object.

_____4. An object's tendency to resist changes in motion.

Part II—Multiple Choice

For each of the following questions, select the best answer.

5. How much force is required to cause a ball with a weight of 3.00 N to accelerate at a rate of 3 m/s^2?

   a) 1 N b) 3 N c) 9 N d) 12 N

6. Forces of 6.0 N and 9.0 N act at right angles to each other on the same box. What is the magnitude of the resultant of these two forces?

   a) 15 N b) 3.0 N c) 11 N d) 1.5 N

7. A box with a mass of 22.5 kg experiences an acceleration of 4.50 m/s^2. Find the net force acting on the object.

   a) 101 N b) 5.00 N c) 18.0 N d) 27.0 N

8. A block of wood with a weight of 5.0 N sits on a horizontal surface. A girl finds that she must apply a force of 2.0 N to overcome the force of friction and get the block to start sliding. What is the coefficient of static friction between the surfaces?

   a) 7.0 b) 3.0 c) 0.40 d) 2.5

9. An ice skater is at rest on the surface of the ice. Compared to the force the skater exerts on the ice, the magnitude of the force the ice exerts on the skater is

   a) greater
   b) less
   c) the same
   d) not enough information given

10. A girl tries to get her dog to start walking by applying a gentle force of 2.8 N to his leash at an angle of 60.0° above the horizontal. What is the magnitude of the horizontal component (F_x) of this applied force?

    a) 1.2 N b) 1.4 N c) 1.8 N d) 2.0 N

11. A desk with a weight 670 N rests on a floor where the coefficient of static friction between the surfaces is 0.45. How much force would be required to get the desk to start sliding?

    a) 3.0×10^2 N
    b) 1.5×10^2 N
    c) 3.1×10^1 N
    d) 3.3×10^3 N

12. A resultant force of 11.0 N is made up of two component forces acting at right angles to each other. If the magnitude of one of the component forces is 6.00 N, what is the magnitude of the other force?

    a) 7.00 N b) 9.22 N c) 11.7 N d) 15.0 N

13. Two students want to work together to push a heavy lab table across the floor. The resultant force will be greatest if the angle between their component forces is

    a) 0° b) 45° c) 90° d) 180°

14. Two forces with magnitudes of 5.0 N and 16.0 N act on the same object at an angle of 180° to each other. What is the magnitude of the resultant force?

    a) 11.0 N b) 15 N c) 17.0 N d) 21 N

Part III—Calculations

Perform the following calculations.

15. A girl pushes a 10.0 N book across a table with a horizontal applied force of 15.0 N. If the coefficient of friction between the book and table is 0.35, find its acceleration.

16. A 5.00 kg wood block slides with uniform velocity down an inclined plane at an angle of 33.0° above the horizontal. Find the coefficient of kinetic friction between the block and the inclined plane.
17. A box with a mass of 22.0 kg is placed on a ramp that is at an angle of 38.0° above the horizontal. What is the magnitude of the component of the weight that is parallel to the surface of the ramp?
18. A woman applies a horizontal force of 50.0 N to push a 200.0 N couch across the floor. If the force of friction between the floor and the couch is 20.0 N, what will be the acceleration of the couch?
19. A man with a weight of 868 N stands on a scale in an elevator that is accelerating upwards at a rate of 2.50 m/s^2. Assuming the scale measures weight in newtons, what will the scale read?
20. The same man from problem 19 is still standing on the scale in the elevator when it starts to accelerate downwards at a rate of 3.80 m/s^2. What will the scale read at this time?

Answer Key

The actual answers will be shown in brackets, followed by an explanation. If you don't understand an explanation that is given in this section, you may want to go back and review the lesson that the question came from.

Lesson 2–1 Review

1. [net force]—You may also see the terms "resultant force" or "total force" used interchangeably with net force.
2. [5.5 N]—The weight of the book and the normal force exerted by the table on the book will add up to 0 N in the vertical (*y*) dimension. In the horizontal, we have an applied force of 12.0 N and a force of friction of 6.5 N. Because friction opposes the motion of the book, I will make it negative as I add the forces.

   $F_{net} = F_a + F_f = 12.0\text{ N} + (-6.5\text{ N}) = 5.5\text{ N}.$
3. [9.43 N]—The question only asks for the magnitude of the force, which means that we don't need to report direction. Because the two component forces act at right angles to each other, we can solve for the resultant force using the Pythagorean theorem.

   $$F_r = \sqrt{F_1^2 + F_2^2} = \sqrt{(5.00\text{ N})^2 + (8.00\text{ N})^2} = 9.43\text{ N}$$

Lesson 2–2 Review

1. [inertia]—To be sure that you understand the concept of inertia, think of some real-life situations where your own inertia is apparent.

Lesson 2–3 Review

1. [weight]—Remember, the formula for weight ($F_w = mg$) is really just a specific application of Newton's second law ($F = ma$).

2. [30.1 kg]—$m = \frac{F_w}{g} = \frac{295\text{ N}}{9.81\text{ m/s}^2} = 30.1\text{ kg}$

3. [2.27 m/s²]—$a = \frac{F_{net}}{m} = \frac{125\text{ N}}{55.0\text{ kg}} = 2.27\text{ m/s}^2$

Lesson 2–4 Review

1. [24.0 N]—Because the question only asks for the magnitude of the force exerted on the rock by the driveway, we don't need to worry about signs. The normal force will be equal to the weight of the rock.

   $F_N = F_w = mg = (2.45\text{ kg})(9.81\text{ m/s}^2) = 24.0\text{ N}$

2. [no]—Newton's third law tells us that the force that the car exerts on the garbage can will be equal and opposite to the force that the garbage can exerts on the car. The two objects have very different masses, so they will experience very different accelerations, as given by Newton's second law:

   $a = \frac{F_{net}}{m}$.

3. [The thin ice is incapable of exerting a force that is equal and opposite to the person's weight, so there is an unbalanced downward (net) force. Therefore, the person accelerates down through the ice.]—The unbalanced downward force before the ice breaks would be equal to the force of weight (F_w) + the force that the ice exerts on the person (F_{ice}). $F_{net} = F_w + F_{ice}$. The F_{ice} would be upward (positive) and the F_w would be downward (negative).

Lesson 2–5 Review

1. [static friction]—*Static* means "unchanging" or "unmoving."
2. [coefficient of friction]—Rough surfaces will have greater coefficients of friction than smooth surfaces.

Chapter 2 Examination

1. [c. friction]—You may have selected "d. contact force," but this definition was more specific. Friction was a better answer because it is a specific type of contact force.
2. [e. weight]—You may have selected "g. field force" but that answer is too general.
3. [b. net force]—The term "resultant force" is sometimes used to refer to the same thing.
4. [h. inertia]—Because of inertia, an object at rest tends to remain at rest and an object in motion tends to remain in motion.
5. [a. 1 N]—We need to use Newton's second law, F = ma, to solve this one. However, notice that they give us the weight of the object, rather than the mass. So we need to use F_w = mg to find the mass of the object first.

   $$m = \frac{F_w}{g} = \frac{3.00 \text{ N}}{9.81 \text{ m/s}^2} = 0.306 \text{ kg}$$

   $$F = ma = (0.306 \text{ kg})(3 \text{ m/s}^2) = 1 \text{ N}$$

6. [c. 11 N]—$F_r = \sqrt{F_1^{\,2} + F_2^{\,2}} = \sqrt{(6.0 \text{ N})^2 + (9.0 \text{ N})^2} = 11 \text{ N}$
7. [a. 101 N]—$F = ma = (22.5 \text{ kg})(4.50 \text{ m/s}^2) = 101 \text{ N}$
8. [c. 0.40]—$\mu_s = \dfrac{F_f}{F_N} = \dfrac{2.0 \text{ N}}{5.0 \text{ N}} = 0.40$
9. [c. the same]—Newton's third law tells us that the forces will be equal and opposite.
10. [b. 1.4 N]—$F_x = F\cos\theta = (2.8 \text{ N})(\cos 60.0^\circ) = 1.4 \text{ N}$
11. [a. 3.0×10^2 N]—To get the desk moving, we must apply a force equal in magnitude to the maximum static friction.

    $$F = F_f = \mu_s F_N = (0.45)(670 \text{ N}) = 301.5 \text{ N} = 3.0 \times 10^2 \text{ N}$$

12. [b. 9.22 N]—The two component forces are acting perpendicular to each other, so the resultant vector of 11.0 N will form the hypotenuse of the right triangle. We can solve this problem with the Pythagorean theorem.

    $$a = \sqrt{c^2 - b^2} = \sqrt{(11.0 \text{ N})^2 - (6.00 \text{ N})^2} = 9.22 \text{ N}$$

13. [a. 0°]—At 45° or 90°, only a component of one force will be added to the other. At 180°, the students will be working against each other.

14. [a. 11.0 N]—If the angle between them is 180°, then they are working against each other. We make one negative and then add them together.
16.0 N + (-5.0 N) = 11.0 N

15. [11.3 m/s²]—We want to use Newton's second law of motion, $F_{net} = ma$, to find the acceleration of the book. In order to do this, we need to find the net force in the *x* dimension. The only forces acting in the *x* dimension are the applied force (F_a) and the force of friction (F_f). We have the applied force, but we don't yet have the force of friction. We do, however, have the coefficient of friction (μ_k) and we can assume that the normal force will be equal in magnitude to the weight of the book (F_w), because the surface is horizontal.

$$\text{force of friction } (F_f) = \mu_k F_N = \mu_k F_w = (0.35)(10.0\text{ N}) = 3.5\text{ N}$$

Remember that the force of friction opposes the applied force, so I will give it a negative sign in the next formula.

$$\text{net force } (F_{net}) = F_a + F_f = 15.0\text{ N} + (-3.5\text{ N}) = 11.5\text{ N}$$

Now we can find the mass of the book in kilograms.

$$\text{mass of book (m)} = \frac{Fw}{g} = \frac{10.0\text{ N}}{9.81\text{ m/s}^2} = 1.02\text{ kg}$$

We now have everything we need to find the acceleration of the book.

$$\text{acceleration of book (a)} = \frac{F_{net}}{m} = \frac{11.5\text{ N}}{1.02\text{ kg}} = 11.3\text{ m/s}^2$$

16. [0.650]—If you don't think that you have enough information to solve this problem, then you probably missed an important "implied given." When we said that the wood block is sliding with "uniform velocity," meaning with an acceleration of zero, we are implying that the net force parallel (F_{netx}) to the surface of the incline must also be zero. Because the only two forces acting parallel to the surface are the force of friction (F_f) and the parallel component of the weight of the block (F_{wx}), they must be equal and opposite in order to add to zero. $F_{netx} = F_{wx} + F_f = 0\text{ N}$

So, our plan should be to find the components of the weight and use the parallel component (F_{wx}) to find the force of friction (F_f) and the perpendicular component (F_{wy}) to find the normal force (F_N). We will then have enough to solve for the coefficient of friction.

$$F_f = -F_{wx} = F_w \sin\theta = -\,mg\sin\theta = -(5.00\text{ kg})(-9.81\text{ m/s}^2)(\sin 33°) = 26.7\text{ N}$$

$$F_N = -F_{wy} = -F_w\cos\theta = -\,mg\cos\theta = (5.00\text{ kg})(-9.81\text{ m/s}^2)(\cos 33°) = 41.1\text{ N}$$

$$\mu_k = \frac{F_f}{F_N} = \frac{26.7\text{ N}}{41.1\text{ N}} = 0.650$$

17. [133 N]—$F_{wx} = mg\sin\theta = (22.0\ \text{kg})(9.81\ \text{m/s}^2)(\sin 38^\circ) = 133\ \text{N}$

18. [1.47 m/s²]—From Newton's second law, we know that $a = \frac{F_{net}}{m}$. We can find the net force by adding the applied force to the force of friction, where we make friction negative. $F_{net} = F_a + F_f = 50.0\ \text{N} + (-20.0\ \text{N}) = 30.0\ \text{N}$
We can find the mass of the couch by dividing its weight by the acceleration due to gravity.

$$m = \frac{F_w}{g} = \frac{200.0\ \text{N}}{9.81\ \text{m/s}^2} = 20.4\ \text{kg}$$

Putting it all together, we get:

$$a = \frac{F_{net}}{m} = \frac{30.0\ \text{N}}{20.4\ \text{kg}} = 1.47\ \text{m/s}^2.$$

19. [1090 N]—It is important to note that the elevator is accelerating. A scale shows the normal force (F_N) that the spring in it is exerting in reaction to the force that is being exerted on it. If the elevator were moving with constant velocity, the man would be exerting a force equal to his weight (F_w) on the scale, and the scale would exert an equal and opposite ($F_N = -F_w$) force. Because the elevator is accelerating, the scale is sort of being pushed into the man's feet. We need to multiply the acceleration of the elevator (a_e) by the mass of the man, and add that to the force of weight.
The mass of the man will be:

$$m = \frac{F_w}{g} = \frac{-868\ \text{N}}{-9.81\ \text{m/s}^2} = 88.5\ \text{kg}.$$

The force shown by the scale will be:

$$F_N = -F_w + ma_e = 868\ \text{N} + (88.5\ \text{kg})(2.50\ \text{m/s}^2) = 1090\ \text{N}.$$

*Note, it is easy to become confused by the sign conventions involved in this type of problem and, in fact, some people might take issue with the way that I used signs in my solution. The important thing to remember is that the forces work together in this problem, resulting in a total force that is greater than the weight of the man. This matches what you experience in real life. As an elevator starts moving (accelerating) upwards, you feel as though you are being pressed into the floor.

20. [532 N]—$F_N = -F_w + ma_e = 868\ \text{N} + (88.5\ \text{kg})(-3.8\ \text{m/s}^2) = 532\ \text{N}$

Work, Energy, Power, and Momentum

Like any other science, physics has its own specific language. Unlike the majority of the terms that you encounter in other sciences, many of the terms that you will encounter in physics are variations of words that you use in everyday life. The advantage to this is that you won't need to learn the spellings of new terms, like, say, *echinoderm*. The disadvantage is that you might have to "unlearn," or at least stop using some previous definitions that you know for terms such as *work*, *power*, and *momentum*. All in all, this should be far less painful than learning to spell *endoplasmic reticulum*.

Lesson 3–1: Work

At some point in your years of studying science, you probably memorized a definition for energy that went something like "Energy is the ability to do work." Unfortunately, as is too often the case in science classes, you may have memorized the definition and never really understood what was meant by "work." You may have assumed that your teacher was using the term work in the same way that it is used in everyday life, but that is not entirely correct. In physics, the term work has a very specific meaning. **Work** is defined as the product of the displacement and the component of the force in the direction of the displacement.

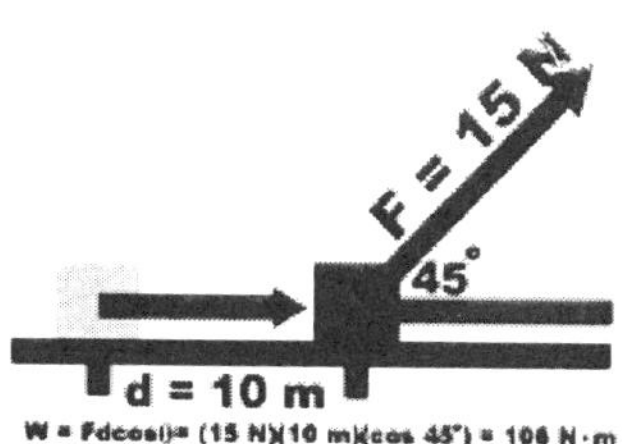

Figure 3.1

Work = force applied in the direction of motion × displacement

$$W = Fd\cos\theta$$

Work is measured in joules (J)

1 joule = N × m = kg × m^2/s^2

It is important to remember that without a displacement, the work done is equal to zero. Holding stationary objects does not count as work. The part of the work equation that shows cos θ also leads to some interesting, yet tricky, questions. The cosine of a 0° angle is equal to 1, so when the displacement and force lie along the same angle, the formula can be simplified to read W = Fd. The cosine of a 90° is equal to 0, so when the applied force and the displacement are at right angles to each other, no work is done by the force. It is also important to note that although work is a scalar quantity, it can have positive or negative values. Work can be done *on* something, or work can be done *by* something.

To get a better understanding of this definition, let's consider some examples.

Example 1

How much work is done by a weight lifter as he holds a 100.0 N barbell stationary, above his head?

The weight lifter is not displacing the weight during this time, so he is doing no work. He did work getting the barbell into its current position, but as he holds it stationary, no work is being done, as shown in the following calculation.

Given: F = 100.0 N d = 0 m $\theta = 0^\circ$

Find: W

Solution: $W = Fd\cos\theta = (100.0\text{ N})(0\text{ m})(\cos 90^\circ) = \mathbf{0\ J}$

Example 2

A lumberjack carries a 50.0 N log on his shoulder as he walks at a constant rate for a distance of 30.0 m across a level surface. How much work does he do on the log during this period of time?

Remember, the lumberjack is walking at a constant rate across a level surface. The log is not accelerating, so there is no net force being exerted on the log along the horizontal axis. The lumberjack is exerting a constant force of 50.0 N on the log with his shoulder, but the direction is upwards, at a right angle to its motion.

Given: F = 50.0 N d = 30.0 m $\theta = 90.0^\circ$

Find: W

Solution: $W = Fd\cos\theta = (50.0\ \text{N})(30.0\ \text{m})(\cos\ 90.0^\circ) = \mathbf{0\ J}$

It is starting to seem like nobody is doing any work. Let's see some more examples.

Example 3

A girl exerts a force of 35.0 N at an angle of 35.0° above the horizontal along the handle of her wagon as she pulls it 25.0 m across a level sidewalk. How much work does she do on her wagon?

Given: F = 35.0 N d = 25.0 m $\theta = 35.0^\circ$

Find: W

Solution: $W = Fd\cos\theta = (35.0\ \text{N})(25.0\ \text{m})(\cos 35.0^\circ) = \mathbf{717\ J}$

Example 4

Attempting to stop a runaway school bus, a superhero exerts a force of 200.0 N against the motion of the bus, slowing it to a stop over a distance of 12.0 m. How much work does the hero do on the bus?

Given: F = 200.0 N d = 12.0 m $\theta = 180^\circ$

Find: W

Solution:

$$W = Fd\cos\theta = (200.0\ \text{N})(12.0\ \text{m})(\cos 180^\circ) = \mathbf{-2.40 \times 10^3\ J}$$

Note the fact that the work done on the bus is negative. This is because the cosine of an 180° angle is equal to –1. The net effect of this work being done on the bus will be to cause a negative acceleration on it. If you are wondering why the answer is in scientific notation, it is because our answer came out to 2400 J, but I wanted to show three significant digits.

Do you know how much work the bus did on the superhero in Example 4? Newton's third law tells us that if the superhero exerted a 200.0 N force on the bus, then the bus exerted a 200.0 N force on the superhero. The superhero was displaced in the same direction as the force applied on him by the bus, so the angle is 0°. The work done by the bus on the superhero would be:

$$W = Fd\cos\theta = (200.0\,\text{N})(12.0\,\text{m})(\cos\ 0^\circ) = \mathbf{2.40 \times 10^3\ J}$$

Sometimes you will get a tricky problem involving gravity. Consider the following example.

Example 5

A warehouse worker pushes an 80.0 kg crate a distance of 3.0 m up a 22.0° incline. How much work is done against gravity?

Be careful with this one. If this question had asked for the total amount of work that the worker did, it would have be equal to

$$W = Fd\cos\theta = mgCos\,\theta = (80.0\,\text{kg})(9.81\,\text{m/s}^2)(3.0\,\text{m})(\cos 0^\circ) = 2400\,\text{J}.$$

However, the question only asks for how much of the work that the worker does is done "against gravity." In other words, we need to find the *y* component of the work. You might be tempted to just change your calculation to include the 22.0° angle, but you need to remember that the angle that we are interested in is always the angle between the displacement and the applied force, and that the question specifically asks for the "work done against gravity." So, the angle that we are interested is actually (90° – 22.0° = 68.0°) 68.0°.

Given: m = 80.0 kg g = 9.81 m/s² d = 3.0 m θ = 68.0°

Find: W

Solution:

$$W = Fd\cos\theta = mgd\cos\theta$$
$$= (80.0\,\text{kg})(9.81\,\text{m/s}^2)(3.0\,\text{m})(\cos 68.0^\circ) = \mathbf{880\ J}$$

Alternatively, we could have used sin 22.0° in place of cos 68.0° and still found the same answer, but I didn't want to confuse anyone by changing the work formula at this point.

Finally, let's do one last example where work isn't the unknown, just so we practice isolating a different variable from the equation.

Example 6

A crane does 680 J of work as it lifts a crate a height of 20.0 m. What is the mass of the crate?

Given: W = 680 J   d = 20.0 m $g = 9.81\ m/s^2$ $\theta = 0^o$

Find: m

Formula: $W = mgd\cos\theta$

Solution:

$$m = \frac{W}{gd\cos\theta} = \frac{680\ \text{J}}{(9.81\ \text{m/s}^2)(20.0\ \text{m})(\cos 0^o)}$$

$$= \frac{680\ \text{kg} \times \cancel{\text{m}}^2/\cancel{\text{s}}^2}{(9.81\ \cancel{\text{m}}/\cancel{\text{s}}^2)(20.0\ \cancel{\text{m}})(1)} = \mathbf{3.5\ kg}$$

Of course, we wouldn't need a crane to pick up the crate from Example 6. A kilogram weighs only about 2.2 pounds near the surface of Earth, so the "crate" weighs only about (3.5 ~~kg~~)(2.2 lb./~~kg~~) = 7.7 lb.!

Now go on to the review questions.

Lesson 3–1 Review

1. ______________ is the product of the displacement and the component of the force in the direction of the displacement.
2. A boy drags a sled 12.0 m across the snow by exerting a 20.0 N force along a rope at an angle of 55.0° above the horizontal. How much work does the boy do on the sled?
3. A worker does 350. J of work lifting a 15.0 kg item to a shelf. How high does she lift it?

Lesson 3–2: Energy

In the last lesson, we introduced work by first defining **energy** as the ability to do work. Now that you understand the concept of work better, you may also have a better understanding of energy. A moving car has energy, and it can do work (exert a force over a distance) on an object that it strikes. A compressed spring has energy, and it can exert a force and displace an object as it expands. A heavy chandelier has energy, and it can fall off the ceiling and do work on objects it lands on, exerting a force and displacing it downwards. Objects with energy can do work, and when we do work on objects, we can give them energy. Work and energy are intimately linked. So much so, in fact, they are both measured in the same units, joules.

Gravitational Potential Energy

Towards the end of our last lesson, we went over an example involving a crane lifting an object. When we lift objects, we use energy to do work on them. Where does the energy that we use go? The energy used gets stored by the object, in a form called **gravitational potential energy**. Gravitational potential energy is the stored energy that an object has due to its own mass, and its position above a reference point.

Gravitiational Potential Energy

The stored energy that an object possesses due to its mass and its height above a reference point.

Gravitational Potential Energy = mass × acceleration due to gravity × height

$$P.E._g = mgh$$

When determining the gravitational energy that an object has, it is important to pay attention to the reference point from which the height is being measured. For example, if you have a can of paint on a shelf and the shelf is on the third floor of a building, you could measure the height of the can with respect to the floor of the room, or to the ground outside the building.

Let's look at an example of a problem that will clearly illustrate the relationship between work and gravitational potential energy.

Example 1

A girl lifts a 5.00 kg toy to a height of 0.75 meters in order to place it on a shelf. Determine (a) how much work (W) the girl does on the toy to lift it, and (b) how much gravitational potential energy ($P.E._g$) the toy has while resting on the shelf, with reference to the floor.

Given: $m = 5.00$ kg $h = 0.75$ m $g = 9.81$ m/s^2

Find: W and P.E.$_g$

Solution:

$$\text{Work done on toy (W)} = Fd = F_w h = mgh$$
$$= (5.00\ \text{kg})(9.81\ \text{m/s}^2)(0.75\ \text{m}) = \mathbf{37\ J}$$

Gravitational Potential Energy of toy

$$(\text{P.E.}_g) = mgh$$
$$= (5.00\ \text{kg})(9.81\ \text{m/s}^2)(0.75\ \text{m})$$
$$= \mathbf{37\ J}$$

As you can see, the girl did 37 joules of work on the toy, and the toy gained 37 joules of gravitational potential energy.

Elastic Potential Energy

Another type of potential energy, called **elastic potential energy**, can be found in springs. When a spring is compressed or stretched, it has the ability to do work in order to restore itself to its equilibrium (rest) position. Have you ever tested the shock absorbers on a car? You push down on the hood of the car, exerting a force. You do work on the car as you exert a force over a distance, pushing the front of the car down. This work gets stored as elastic potential energy, which is used to restore the car to its normal position, after you stop applying a force.

Another example of this type of energy is found when you shoot a rubber band. You stretch a rubber band over one of your fingers. As you stretch the rubber band, you are doing work on it, exerting a force over a distance. The energy that you expend is stored in the rubber band as elastic potential energy. When you release the rubber band, this energy is converted to kinetic energy, causing the band to launch toward your intended victim.

Different rubber bands and springs require different amounts of force to stretch or compress them. In each case, however, the distance that they are compressed or stretched is proportional to the force applied. The ratio of the force applied to a spring and the distance it is stretched or compressed gives us the proportionality constant (k), which is appropriately called the **spring constant**. The "stiffer" a spring is, the more force is required to stretch or compress it by a certain distance, the higher the value is for k. We can determine the spring constant for a spring by determining how much it will be stretched by a certain force.

If we know the spring constant of a spring, we can use a relationship called Hooke's law to determine the force the spring will exert when it has been displaced by a certain distance.

Hooke's Law

Elastic force = –(spring constant × displacement)

$$F_{elas} = -kx$$

Example 2

A mass of 5.0 kg is attached to a hanging vertical spring. The spring stretches 4.0 cm beyond its equilibrium position before coming to a rest. Determine the spring constant for the spring.

Figure 3.2

Convert:

First, let's convert centimeters to meters.

$$4.0\ \cancel{cm} \times \frac{1\ m}{100\ \cancel{cm}} = 0.040\ m$$

Now we find the displacement of the mass.

$x = x_f - x_i = 0.00\ m - 0.040\ m = \mathbf{-0.040\ m}$

Let us also find the weight of the object on the spring, which will be the force applied.

$$F_w = mg = (5.0\ kg)(-9.81\ m/s^2) = 49.05\ N = \mathbf{-49\ N}$$

Because the mass is at rest, we know that the elastic force (F_{elas}), the force being exerted by the spring on the mass, must be equal and opposite to the force of weight.

$F_{net} = F_w + F_{elas} = -49\ N + 49\ N = 0\ N$

Given: $F_{elas} = 49\ N$ $\qquad$ $x = -0.040\ m$

Find: k

Isolate:

Starting with: $F_{elas} = -kx$

We simply divide both sides by –x: $\frac{F_{elas}}{-x} = \frac{\cancel{-}k\cancel{x}}{\cancel{-}\cancel{x}}$

Rearranging, we get: $k = \frac{F_{elas}}{-x}$

Solution: $k = \frac{F_{elas}}{-x} = \frac{49\text{ N}}{0.040\text{ m}} = 1225\text{ N/m} = \mathbf{1200\text{ N/m}}$

This is a very high spring constant, indicating a very stiff spring. This makes sense for our example because a mass of 5.0 kg (weighing about 11 pounds) only stretched this imaginary spring 4.0 centimeters.

Example 3

How much force would a spring with a spring constant of 22 N/m exert when it has been compressed 0.25 m from its equilibrium (rest) position?

Given: k = 22 N/m x = 0.25 m

Find: F_{elas}

Solution: $F_{elas} = -kx = -(22\text{ N/}\cancel{m})(0.25\text{ }\cancel{m}) = \mathbf{-5.5\text{ N}}$

The negative sign in our answer simply indicates that the spring will exert a force in the direction opposite to the displacement. For this reason, you will sometimes see this elastic force referred to as a *restorative* force.

The amount of elastic potential energy ($P.E._{elas}$) stored in a spring can be determined with the following formula.

$$P.E._{elas} = \frac{1}{2}kx^2$$

Example 4

Calculate the elastic potential energy stored in a spring with a spring constant of 45 N/m that has been compressed 1.2 m from its equilibrium position.

Given: k = 45 N/m x = 1.2 m

Find: $P.E._{elas}$

Solution: $P.E._{elas} = \frac{1}{2}kx^2 = \left(\frac{1}{2}\right)(45\ N/m)(1.2\ m)^2 = 32.4\ J = \mathbf{32\ J}$

Kinetic Energy

Work doesn't always increase the potential energy of an object. For example, if you do work on a hockey puck horizontal to the surface of the ice, you won't increase the height or gravitational potential energy of the object. You will, however, increase the kinetic energy of the hockey puck. The **kinetic energy** is the energy that an object has due to its motion. Like work and gravitational potential energy, kinetic energy is measured in joules.

Kinetic Energy

The energy of an object due to its motion.

$$K.E. = \frac{1}{2}mv^2$$

Example 5

Calculate the kinetic energy of a rolling ball with a mass of 5.00 kg and a velocity of 2.50 m/s.

Given: m = 5.00 kg $v = 2.50\ m/s$

Find: K.E.

Solution: $K.E. = \frac{1}{2}mv^2 = (0.5)(5.00\ kg)(2.50\ m/s)^2 = \mathbf{15.6\ J}$

Example 6

A toy car with a mass of 3.5 kg has 63 J of kinetic energy as it drives along the straight part of a racetrack. Calculate the velocity of the toy car.

Given: K.E. = 63 J m = 3.5 kg

Find: v

Isolate:

Starting with: $K.E. = \frac{1}{2}mv^2$

Multiply both sides by 2: $2K.E. = 2 \times \frac{1}{2}mv^2$

We get: $2K.E. = mv^2$

Now, divide both sides by m: $\frac{2\,K.E.}{m} = \frac{\not{m}v^2}{\not{m}}$

Rearrange and take the square root of both sides: $\sqrt{v^2} = \sqrt{\frac{2\,K.E.}{m}}$

We get: $v = \sqrt{\frac{2\,K.E.}{m}}$

Solution: $v = \sqrt{\frac{2\,K.E.}{m}} = \sqrt{\frac{2(63\text{ J})}{3.5\text{ kg}}} = \mathbf{6.0\ m/s}$

Lesson 3–2 Review

1. ________________ is the energy that an object possesses due to its motion.
2. Calculate the elastic potential energy stored in a spring with a spring constant of 25 N/m that has been compressed 0.874 m from its equilibrium position.
3. Calculate the kinetic energy of a 1.20×10^3 kg car with a velocity of 35.0 m/s.

Lesson 3–3: Conservation of Energy

One of the most important laws of physics is the law of conservation of mechanical energy, which states that in the absence of dissipative forces (such as friction), the total amount of mechanical energy in a system is conserved. In physics we define **mechanical energy** as the sum of the kinetic energy and all forms of potential energy in a system. When an object falls, its gravitational potential energy is converted into kinetic energy, but the total mechanical energy (when we ignore friction) remains the same. It may seem strange to you that we are ignoring friction once again. If we did a problem where an object was falling from such a great height that it reached terminal velocity, it certainly would be inappropriate to ignore friction, because the object would clearly be losing gravitational potential energy without gaining more kinetic energy. However, the problems that we will be dealing with involve small falls and therefore insignificant losses due to friction.

Example 1

A 5.4 kg bowling ball falls from rest off a shelf at a height of 1.2 m. Calculate both the gravitational potential energy of the ball with respect to the floor and the kinetic energy of the ball (a) just before it falls, when its height is 1.2 m; (b) when it is falling and its height is 0.60 m; and (c) the instant before the ball hits the ground, when its height would round to 0.0 m above the ground.

This represents a classic problem used to show conservation of mechanical energy.

a) **At a height of 1.2 m.**

When the ball is at its maximum height, its gravitational potential energy will be

$$P.E._g = mgh = (5.4\text{ kg})(9.81\text{ m/s}^2)(1.2\text{ m}) = \mathbf{64\ J}.$$

Because the ball hasn't started falling at this point, its velocity is zero, and its kinetic energy will be equal to

$$K.E. = \frac{1}{2}mv^2 = \left(\frac{1}{2}\right)(5.4\text{ kg})(0.0\text{ m/s})^2 = \mathbf{0\ J}.$$

The total of these two forms of energy must remain the same for each part of the problem.

$$E_{total} = P.E.g + K.E. = mgh + \frac{1}{2}mv^2 = 64\text{ J} + 0\text{ J} = \mathbf{64\ J}$$

Knowing this will save us a good deal of work while solving the rest of the problem.

b) At a height of 0.60 m.

When the ball has fallen halfway and its height has decreased to 0.60 m, its gravitational potential energy will be

$P.E._g = mgh = (5.4 \text{ kg})(9.81 \text{ m/s}^2)(0.60 \text{ m}) = \mathbf{32\ J}$.

We can easily find the kinetic energy at this height because the total mechanical energy must be conserved.

$K.E. = E_{total} - P.E.g = 64 \text{ J} - 32 \text{ J} = \mathbf{32\ J}$

If we didn't know this relationship, we would have to use a formula for free fall to find the velocity of the bowling ball at this height, as shown here:

$$v_f = \sqrt{2gd} = \sqrt{2(9.81 \text{ m/s}^2)(0.60 \text{ m})} = \mathbf{3.43\ m/s}^*.$$

(* I kept an extra digit in my answer to velocity to avoid a rounded answer in the kinetic energy that would look slightly different than the kinetic energy.)

Using this velocity, we could find the kinetic energy of the bowling ball at 0.60 m.

$$K.E. = \frac{1}{2}mv^2 = \left(\frac{1}{2}\right)(5.4 \text{ kg})(3.43 \text{ m/s})^2 = \mathbf{32\ J}$$

Just looking at the formula for velocity and substituting it into our formula for kinetic energy, you can see that it will be equal to our formula for gravitational potential energy.

$$K.E. = \frac{1}{2}mv^2 = \frac{1}{2}m(\sqrt{2gd})^2 = \frac{1}{2}m(2gd) = mgd = mg\Delta h = \mathbf{\Delta P.E.g}$$

c) At a height of 0.0 m.

Finally, we can find the gravitational potential energy and the kinetic energy of the bowling ball the instant before it hits the ground. The key here is to remember that the bowling ball is still moving (velocity is not zero) but the height is essentially zero, right before it strikes the ground. At this point, all of the original gravitational potential energy has been converted to kinetic energy.

$P.E._g = mgh = (5.4 \text{ kg})(9.81 \text{ m/s}^2)(0.0 \text{ m}) = \mathbf{0\ J}$

$K.E. = E_{total} - P.E.g = 64 \text{ J} - 0 \text{ J} = \mathbf{64\ J}$

It took us a long time to get here, but hopefully you see that these problems are actually quite easy. All you need to do is find the original gravitational potential energy and add or subtract from there. It may help to see the problem summarized in the following table.

Height	Gravitational Potential	Kinetic Energy	Total Mechanical Energy
	$P.E._g = mgh$	$K.E. = \frac{1}{2}mv^2$	$E_{total} = P.E._g + K.E.$
1.2 m	64 J	0 J	64 J
0.60 m	32 J	32 J	64 J
0.0 m	0 J	64 J	64 J

The relationship between gravitational potential energy and kinetic energy can be used to solve another type of interesting problem, as shown in our next example.

Example 2

A child slides down a frictionless slide from a height of 1.8 m. How fast will she be going at the bottom of the slide?

Given: $g = 9.81 \text{ m/s}^2$ $\qquad h = 1.8 \text{ m}$

Find: v

Now, this is where a novice student begins to panic. You need to find the speed of the child at the bottom of the slide, but you only have the height of the slide and the value of g. You can't think of any formula that will allow you to solve for velocity with only this given information. However, if you remember that the kinetic energy that the girl gains is equal to the gravitational potential energy that she loses, you can set the two equations equal to each other and solve for velocity.

Isolate:

$$P.E._g = K.E. \text{ so, } mgh = \frac{1}{2}mv^2$$

Multiply both sides by 2: $2mgh = 2 \times \frac{1}{2}mv^2$

We get: $2mgh = mv^2$

Now, divide both sides by m: $\frac{2\not{m}gh}{\not{m}} = \frac{\not{m}v^2}{\not{m}}$

We get: $2gh = v^2$

Rearrange and take the square root of both sides to get: $v = \sqrt{2gh}$

Solution: $v = \sqrt{2gh} = \sqrt{2(9.81\,\text{m/s}^2)(1.8\,\text{m})} = \mathbf{5.9\,m/s}$

This example reminds us that we don't need to know the mass of an object to determine its final velocity, as the speed of a falling object is not dependent upon its mass.

Lesson 3–3 Review

1. A box with a mass of 12.0 kg slides down a frictionless slide with a height of 3.00 m. What will be the kinetic energy of the box at the bottom of the ramp?
2. An apple with a mass of 0.10 kg falls off a tree branch with a height of 2.75 m. Find the velocity of the apple when it hits the ground.
3. A bowling ball with a mass of 5.00 kg falls off a shelf with a height of 2.00 m. How much kinetic energy would the bowling ball have when it has fallen halfway to the floor, when its height is 1.00 m?

Lesson 3–4: Power

Imagine two men employed to restock the shelves of a supermarket with cans of soup. Each time one of the men lifts a can, he exerts a force on it and lifts it up. By exerting a force on the can over a distance, he does work on it. If each man lifts the same number of cans from a box and places them on the same shelf, they have done the same amount of total work. If one man works faster than the other, he will have a higher power rating. **Power** is the rate at which work is done, and it is measured in the units called **watts** (W), which you may be familiar with from your experience changing lightbulbs. One watt is equal to one joule/second. A 60-watt bulb converts 60 joules of electrical energy to heat and light per second.

Power

$$\text{power} = \frac{\text{work}}{\text{change in time}} = \frac{\text{force} \times \text{displacement}}{\text{change in time}} = \text{force} \times \text{velocity}$$

$$P = \frac{W}{\Delta t} = \frac{fd}{\Delta t} = fv$$

Let's go over some examples.

Example 1

A man restocking a shelf lifts 24 cans of soup, each with a mass of 450 g, to a height of 55 cm in a period of 45 seconds. Find his average power output during this period of time.

The first thing that you will need to do is convert all of the given information into appropriate units. Remember, the unit of power (watt) is based on the newton, which is derived from meters and kilograms. We must convert our given quantities, as shown here:

Convert:

$$55\ \cancel{\text{cm}} \times \frac{1\ \text{m}}{100\ \cancel{\text{cm}}} = \mathbf{0.55\ m} \qquad 450\ \cancel{\text{g}} \times \frac{1\ \text{kg}}{1000\ \cancel{\text{g}}} = \mathbf{0.45\ kg}$$

Now let's calculate the total mass of cans that the man lifted.

$$m_{total} = 0.45\ \text{kg}/\cancel{\text{can}} \times 24\ \cancel{\text{can}} = \mathbf{11\ kg}$$

Given: $m = 11$ kg $\quad g = 9.81\ \text{m/s}^2$ $\quad h = 0.55$ m $\quad \Delta t = 45$ s

Find: P

Solution: $$P = \frac{W}{\Delta t} = \frac{mgh}{\Delta t} = \frac{(11\ \text{kg})(9.81\ \text{m/s}^2)(0.55\ \text{m})}{45\ \text{s}} = \mathbf{1.3\ W}$$

Let's try the same situation but with a different unknown.

Example 2

Another man working on restocking the soup from Example 1 had an average power output of 1.6 W lifting 24 cans of soup to the same shelf, which has a height of 0.55 m. How long did it take him to finish the task?

Given: P = 1.6 W m = 11 kg g = 9.81 m/s^2 h = 0.55 m

Find: Δt

Isolate:

Starting with: $P = \dfrac{mgh}{\Delta t}$

Multiply both sides by Δt: $\Delta t \times P = \dfrac{mgh}{\cancel{\Delta t}} \times \cancel{\Delta t}$

And divide both sides by P: $\dfrac{\Delta t\cancel{P}}{\cancel{P}} = \dfrac{mgh}{P}$

Solution: $\Delta t = \dfrac{mgh}{P} = \dfrac{(11\,\text{kg})(9.81\,\text{m/s}^2)(0.55\,\text{m})}{1.6\,\text{W}} = \mathbf{37\,s}$

Another unit of power that you will sometimes encounter is horsepower, or hp for short. One horsepower is equal to 746 watts. If you encounter a problem involving horsepower, convert to watts before trying to solve.

Example 3

A motor has a power rating of 5.5 hp. How much work can it do, operating at full power, in 3.0 minutes?

Convert:

$5.5\,\cancel{\text{hp}} \times \dfrac{746\,\text{W}}{\cancel{\text{hp}}} = 4103\,\text{W} = \mathbf{4100\,W}$ $3.0\,\cancel{\text{min}} \times \dfrac{60\,\text{s}}{\cancel{\text{min}}} = \mathbf{180\,s}$

Given: P = 4100 W Δt = 180 s

Find: W

Isolate:

Multiply both sides of the power formula by Δt:

$P \times \Delta t = \dfrac{W}{\cancel{\Delta t}} \times \cancel{\Delta t}$

Solution: $W = P\Delta t = (4100\,\text{W})(180\,\text{s}) = (4100\,\text{J}/\cancel{\text{s}})(180\,\cancel{\text{s}})$
$= 738\,000\,\text{J} = \mathbf{740\,000\,J}$

Example 4

A worker pushes a box with a force of 35 N over a period of 6.0 s, moving the box 2.5 m. How much power does the worker supply?

Given: F = 35 N Δt = 6.0 s d = 2.5 m

Find: P

Solution: $P = \frac{W}{\Delta t} = \frac{Fd}{\Delta t} = \frac{(35\,\text{N})(2.5\,\text{m})}{6.0\,\text{s}} = 14.5833\bar{3}\ \text{W} = \mathbf{15\ W}$

Lesson 3–4 Review

1. _______________ is the rate of doing work, measured in watts.
2. A crane lifts a packing crate with a mass of 355 kg to a height of 15.5 m with a motor operating at a constant rate of 2.50×10^3 watts. How long does it take to lift the crate?
3. What power would be required to lift a 340 N weight to a height of 3.50 m in 5.0 seconds?

Lesson 3–5: Linear Momentum

Watching a football game on television, you are likely to hear the sportscasters use the term *momentum* in a couple of different ways. Sometimes they talk about which team has *momentum*, referring to which team seems to be controlling the game at the time. They might also talk about the *momentum* of a player carrying him forward. The second use of this term is closer to how we define linear momentum in physics. **Linear momentum** is the product of an object's mass and velocity. We will use the symbol p for momentum, and the units are often shown as kg · m/s.

Linear momentum = mass × velocity

$$p = mv$$

Linear momentum (p) is measured in kg · m/s

For the rest of this lesson, we will simply refer to linear momentum as momentum, which is a common practice.

Momentum is a vector quantity, and the direction of the momentum will be in the same direction as the velocity. As with the case of velocity, the direction of the momentum can be indicated by a compass direction, or by a sign. A negative momentum simply means a momentum in the direction that is considered negative.

Example 1

Calculate the momentum of a 5.5 kg bowling ball traveling with a velocity of –15 m/s.

Given: m = 5.5 kg $\qquad v = -15$ m/s

Find: p

Solution: $p = mv = (5.5\ \text{kg})(-15\ \text{m/s}) = -82.5\ \text{kg}\cdot\text{m/s} = \mathbf{-83\ kg\cdot m/s}$

The momentum of an object will change anytime its velocity (or mass) changes. The tricky part about change in momentum problems is to pay attention to the direction of the velocity. If a ball hits a wall and bounces off at the exact same speed, its speed hasn't changed, but its direction, and therefore its velocity, has. For questions that ask you to calculate the change in the momentum of a single object, where the mass of the object does not change, use the following formula:

$$\Delta p = p_f - p_i = mv_f - mv_i = m(v_f - v_i).$$

Example 2

A baseball with a mass of 0.10 kg has a speed of 20.0 m/s as it flies toward a batter. The batter hits the ball and sends it in the exact opposite direction with a speed of 22.0 m/s. Calculate the change in the ball's momentum.

The key to the problem is to notice that they give you the speed of the ball, not the velocity. If you read the question too fast, you might think that the velocity of the ball has only changed by 2.0 m/s, but that is not the case. Reading the problem more carefully, you find that the batter hit the ball and sent it in the exact opposite direction. That means that we need to make the final velocity –22.0 m/s, and the velocity of the ball actually changed by (–22.0 m/s – 20.0 m/s) –42.0 m/s!

Given: m = 0.10 kg $\qquad v_i = 20.0$ m/s $\qquad v_f = -22.0$ m/s

Find: Δp

Solution:

$$\Delta p = m(v_f - v_i) = 0.10\text{ kg}(-22.0\text{ m/s} - 20.0\text{ m/s})$$
$$= 0.10\text{kg}(-42.0\text{ m/s}) = \mathbf{-4.2\text{ kg}\cdot\text{m/s}}$$

In Lesson 3–2 we discussed the law of conservation of mechanical energy, and said that there are cases in which the total mechanical energy of a system is conserved. There are also situations in which the total momentum of a system is conserved, resulting in the law of conservation of linear momentum.

Law of Conservation of Momentum

The total momentum of a system remains the same when the net external force acting on a system is zero.

General formula for two objects:

$$\Sigma\Delta p = (m_1 v_{1f} + m_2 v_{2f}) - (m_1 v_{1i} + m_2 v_{2i}) = 0$$

This formula is usually shown as $\mathbf{m_1 v_{1i} + m_2 v_{2i} = m_1 v_{1f} + m_2 v_{2f}}$

The momentum of the baseball in Example 2 wasn't conserved because it was acted on by an external force, which was provided by the bat. We could have studied both the bat and the ball as one system, and the law of conservation of momentum would have applied. As long as the air resistance and loss of energy to sound were treated as negligible, we would have found that the momentum gained by the ball was equal to the momentum lost by the bat.

A common external force that prevents momentum from being conserved in many real-life situations is friction. In a game of billiards, there is friction between the balls and the table, and between the balls and the air. If there were no external forces at work in a game of billiards, every break shot, which sets all of the balls in motion, would likely result in each of the balls eventually being pocketed. The balls would simply bounce around on the table, colliding with each other, until they found a pocket.

An air hockey table comes much closer to approaching a frictionless surface. The puck floats on a cushion of air, and it will ricochet around on the table for a good amount of time. However, there is still a small amount of friction between the puck and the air, and a certain amount of energy is converted to sound when the puck hits the sides of the table.

Collisions come in many varieties. When a tennis ball hits a wall, it deforms momentarily, but it springs back into shape as is bounces off the wall. When a car hits a wall, the car deforms quite a bit, and it doesn't spring back into shape, although it may "bounce" back from the wall. If a ball of clay strikes the same wall, it is likely to deform and stick to the wall.

Elastic Collisions

In an isolated **elastic collision**, both the momentum and the kinetic energy are conserved. In other words, the total momentum before the collision is equal to the total momentum after the collision and the total kinetic energy before the collision is equal to the total kinetic energy after the collision. There are no perfectly elastic collisions, but when the loss of kinetic energy is considered insignificant, the collision is still treated as elastic. If you studied the ideal gas laws in chemistry, you may recall that the collisions between molecules of gases are treated as if they are totally elastic. The collisions of a puck on the air hockey table would likely be considered elastic. The example of the tennis ball hitting a wall could be as well.

Many of the problems that you will encounter for momentum will take place on ice or an air table, and they should be treated as isolated systems, where momentum will be conserved. If the objects bounce off each other during collisions, they should be treated as elastic collisions.

One common example of a problem designed to test conservation of momentum involves two ice-skaters facing each other on a frictionless surface of ice. The skaters are initially at rest, so their total momentum before they push off ($m_1v_{1i} + m_2v_{2i} = 0$) is zero. They push off each other, and you are asked to use the mass of the skaters and the final velocity of one of the skaters to find the final velocity of the other. Remember, if the total momentum before and after the push is conserved ($m_1v_{1i} + m_2v_{2i} = m_1v_{1f} + m_2v_{2f}$), then the total momentum after the push ($m_1v_{1f} + m_2v_{2f} = 0$) must also be equal to zero.

Example 3

Two ice-skaters, with masses of 45 kg and 65 kg, respectively, are initially at rest and facing each other over a surface of ice. They push off from each other and the 45 kg skater moves away with a velocity of 6.0 m/s. Find the final velocity of the 65 kg skater.

Given: $m_1 = 45$ kg $m_2 = 65$ kg $v_{1f} = 6.0$ m/s $p_i = 0$ $p_f = 0$

Find: v_{2f}

Isolate:

Initial Formula: $m_1v_{1f} + m_2v_{2f} = 0$

Subtract m_1v_{1f} from both sides: $m_2v_{2f} = -(m_1v_{1f})$

Divide both sides by m_2: $v_{2f} = -\left(\frac{m_1v_{1f}}{m_2}\right)$

Solution:

$$v_{2f} = -\left(\frac{m_1v_{1f}}{m_2}\right) = -\left(\frac{45\text{ kg}\cdot 6.0\text{ m/s}}{65\text{ kg}}\right)$$

$$= -4.1538461\text{ m/s} = \mathbf{-4.2\text{ m/s}}$$

The negative sign in our answer indicates that the 65 kg skater will move in the opposite direction of the 45 kg skater.

As you can see, when the objects start at rest, or when they end at rest, the formula for solving these equations is not very complicated. Let's look at an example where the objects are moving before and after the collision.

Example 4

A ball with a mass of 0.25 kg and an initial velocity of 1.45 m/s has a head-on elastic collision with a 0.45 kg ball with an initial velocity of –2.25 m/s. After the collision, the 0.45 kg ball has a velocity of –0.84 m/s. Calculate the final velocity of the 0.25 kg ball. (Assume the momentum is conserved.)

Given: $m_1 = 0.25$ kg $m_2 = 0.45$ kg $v_{1i} = 1.45$ m/s
$v_{2i} = -2.25$ m/s $v_{2f} = -0.84$ m/s

Find: v_{1f}

Solution: Our solution is based on the formula:

$m_1v_{1i} + m_2v_{2i} = m_1v_{1f} + m_2v_{2f}$

Let's start by calculating the total momentum before the collision:

$$m_1v_{1i} + m_2v_{2i} = (0.25\text{ kg})(1.45\text{ m/s}) + (0.45\text{ kg})(-2.25\text{ m/s})$$
$$= -0.65\text{ kg}\cdot\text{m/s}$$

The law of conservation of momentum tells us that we must have the same momentum after the collision.

$m_1v_{1f} + m_2v_{2f} = -0.65\text{ kg}\cdot\text{m/s}$

Isolating this formula for our unknown (v_{1f}), we get:

$$v_{1f} = \frac{(-0.65\ \text{kg} \cdot \text{m/s}) - m_2 v_{2f}}{m_1}$$

$$= \frac{(-0.65\ \text{kg} \cdot \text{m/s}) - (0.45\ \text{kg})(-0.84\ \text{m/s})}{0.25\ \text{kg}} = \mathbf{-1.1\ m/s}$$

Check: With these problems, I always feel that it is a good idea to check your answer. Let's make sure that the total momentum was conserved by putting our numbers back into the original equation.

$$m_1 v_{1f} + m_2 v_{2f} = (0.25\ \text{kg})(-1.1\ \text{m/s}) + (0.45\ \text{kg})(-0.84\ \text{m/s})$$
$$= -0.65\ \text{kg} \cdot \text{m/s}$$

Our answer is correct, as momentum is conserved. The final velocity of the second ball is –1.1 m/s. It is interesting to note that the ball changed directions after being struck by the more massive, faster moving ball.

Inelastic Collisions

In a totally **inelastic collision**, the total kinetic energy of the system is not conserved. An inelastic collision is sometimes referred to as a "sticky" collision, because the objects remain in contact after the collision. Consider the following example.

Example 5

A block with a mass of 8.0 kg slides across a frictionless surface with an initial velocity of 5.0 m/s. It collides with a 4.0 kg block with a velocity of –10.0 m/s. After the collision, the two blocks stick together. Calculate the final velocity of the blocks.

Given: $m_1 = 8.0$ kg $m_2 = 4.0$ kg $v_{1i} = 5.0$ m/s $v_{2i} = -10.0$ m/s

Find: v_f

Solution:

First, let's find the initial total momentum of the two-block system.

$$p_{total} = m_1 v_{1i} + m_2 v_{2i} = (8.0\ \text{kg})(5.0\ \text{m/s}) + (4.0\ \text{kg})(-10.0\ \text{m/s})$$
$$= 40.\ \text{kg} \cdot \text{m/s} + (-40.\ \text{kg} \cdot \text{m/s}) = \mathbf{0.0\ kg \cdot m/s}$$

The law of conservation of momentum tells us that the total initial momentum must equal the total final momentum, so the final momentum of the two-block system must also be equal to zero. The total mass of the blocks doesn't change to zero just because they collide, so the velocity of the two blocks must equal zero.

You should see that the kinetic energy of this system is not conserved. Clearly the blocks had kinetic energy before they collided, but it is zero after the collision because both blocks end up at rest.

Example 6

A block with a mass of 6.0 kg and an initial velocity of 5.0 m/s collides with a stationary block with a mass of 4.0 kg on a frictionless surface. After the collision, the blocks stay together. What will be the final velocity of the two-block system?

Given: $m_1 = 6.0$ kg $m_2 = 4.0$ kg $v_{1i} = 5.0$ m/s $v_{2i} = 0.0$ m/s
Find: v_f
Solution: The key to solving this problem is remembering that the blocks remain together after the collision, so that the final velocity of the blocks will be the same.

Momentum is conserved, so $p_i = p_f$.

Original formula: $m_1 v_{1i} + m_2 v_{2i} = m_1 v_{1f} + m_2 v_{2f}$

The final velocity of both blocks is the same, so

$m_1 v_{1i} + m_2 v_{2i} = m_1 v_f + m_2 v_f$.

Factoring out v_f: $m_1 v_{1i} + m_2 v_{2i} = v_f(m_1 + m_2)$

Isolating v_f we get: $v_f = \dfrac{m_1 v_{1i} + m_2 v_{2i}}{(m_1 + m_2)}$

$$v_f = \frac{m_1 v_{1i} + m_2 v_{2i}}{(m_1 + m_2)} = \frac{(6.0\ \text{kg})(5.0\ \text{m/s}) + (4.0\ \text{kg})(0.0\ \text{m/s})}{(6.0\ \text{kg} + 4.0\ \text{kg})}$$

$$= \frac{30.\ \cancel{\text{kg}} \cdot \text{m/s}}{10.\ \cancel{\text{kg}}} = \mathbf{3.0\ m/s}$$

Lesson 3–5 Review

1. ________________ is a collision in which the total kinetic energy is conserved.
2. ________________ is the product of the mass and velocity of an object.
3. What is the velocity of a 0.10 kg baseball when its momentum is 3.50 kg · m/s?

Lesson 3–6: Impulse

If you have ever been coached in a sport that involves striking a ball with some type of club, bat, or racket, you have probably been told about the importance of "follow-through." Using tennis as an example, the purpose of follow-through is to increase the amount of time that the ball is in contact with the racket. Remember, the racket is what is applying the force to the tennis ball, so follow-through allows the force to be applied for a longer period of time, resulting in greater acceleration. In physics, the product of the force and the time for which the force is applied is called **impulse**. Impulse is typically represented with the symbol J and is measured in newtons · seconds (N · s).

$$J = F\Delta t$$

The impulse-momentum theorem states that the impulse applied to an object will result in a change in the object's momentum, giving rise to the very useful formulas:

$$F\Delta t = \Delta p \quad \text{or} \quad F\Delta t = m\Delta v \quad \text{or} \quad F\Delta t = m(v_f - v_i).$$

Example 1

A golf club exerts a uniform force of 22.0 N on a golf ball for a period of 0.250 s. Calculate the change in the ball's momentum.

Given: F = 22.0 N Δt = 0.250 s
Find: Δp
Solution:

$$\Delta p = F\Delta t = (22.0\text{ N})(0.250\text{ s}) = (22.0\text{ kg}\cdot\text{m/s}^2)(0.250\text{ s}) = \mathbf{5.50\ kg\cdot m/s}$$

Example 2

A girl playing T-ball hits a 0.20 kg baseball off a tee, exerting a force of 3.5 N for a period of 0.15 s. Find the final velocity of the baseball.

Given: F = 3.5 N m = 0.20 kg Δt = 0.15 s $v_i = 0.0$ m/s

Find: v_f

Isolate:

Starting with: $F\Delta t = m(v_f - v_i)$

We can cross out v_i, because it is equal to zero: $F\Delta t = m(v_f - \cancel{v_i})$

Leaving us with: $F\Delta t = mv_f$

Divide both sides by m: $\frac{F\Delta t}{m} = \frac{\cancel{m}v_f}{\cancel{m}}$

Rearranging the equation, we get: $v_f = \frac{F\Delta t}{m}$

Solution: $v_f = \frac{F\Delta t}{m} = \frac{(3.5\,N)(0.15\,s)}{0.20\,kg} = \mathbf{2.6\,m/s}$

If we rearrange the impulse formula and isolate the force, we can see how it relates to many real-life situations.

$$F = \frac{\Delta p}{\Delta t} = \frac{m\Delta v}{\Delta t} = \frac{mv_f - mv_i}{\Delta t} = \frac{m(v_f - v_i)}{\Delta t}$$

Notice that the instantaneous force that an object experiences is inversely proportional to the amount of time that the force acts. That means that if it takes longer for an object to come to a stop, it will experience a smaller instantaneous force. When a glass drops on a rug, the rug increases the stopping time, preventing the glass from breaking. When you catch a baseball with a glove, you probably move the glove back as you catch the ball, increasing the stopping time and decreasing the instantaneous force.

Example 3

A boy catches a 0.10 kg baseball with an initial velocity of 12.0 m/s, bringing it to rest in 0.89 s. Calculate the average force exerted on the ball during the period it was being brought to rest.

Given: m = 0.10 kg $\quad v_i = 12.0$ m/s $\quad v_f = 0.0$ m/s $\quad \Delta t = 0.89$ s

Find: F

Solution: $F = \frac{m(v_f - v_i)}{\Delta t} = \frac{0.10\ \text{kg}(0.0\ \text{m/s} - 12.0\ \text{m/s})}{0.89\ \text{s}} = \mathbf{-1.3\ N}$

Don't be surprised by the negative sign in our answer, which simply indicates that the boy must exert a force in the opposite direction to stop the ball.

Lesson 3–6 Review

1. ______________ is the product of a force and the time interval during which the force acts.
2. Calculate the impulse of a 20.0 N force acting for 15.0 s.
3. A 0.100 kg puck has an initial velocity of 3.50 m/s until a player hits it with a stick, exerting a force of 12.0 N in the direction of its motion for 0.344 seconds. What is the final velocity of the puck?

Chapter 3 Examination

Part I—Matching

Match the following terms to the definitions that follow.

a. work	d. gravitational potential energy	g. power
b. energy	e. elastic potential energy	h. momentum
c. kinetic energy	f. mechanical energy	i. impulse

_____1. The rate at which work is done.

_____2. The sum of the kinetic and all the forms of potential energy.

_____3. The product of an object's mass and velocity.

_____4. The energy that an object possesses due to its motion.

_____5. The product of the force and the period of time during which the force is acting.

Part II—Multiple Choice

For each of the following questions, select the best answer.

6. Which of the following are units of power?
   a) joules b) watts c) newtons d) $kg \cdot m/s^2$
7. If the velocity of a moving car is doubled, its kinetic energy will be
   a) halved b) doubled c) tripled d) quadrupled
8. A glass has 0.500 J of gravitational potential energy when sitting on a shelf with a height of 0.400 m. What is the mass of the glass?
   a) 0.127 kg b) 1.25 kg c) 3.92 kg d) 0.200 kg
9. An apple has 0.800 J of gravitational potential energy when hanging from a tree. If it falls off the branch, how much gravitational potential energy will it have when it has fallen half the distance to the ground?
   a) 1.60 J b) 0.400 J c) 3.92 J d) 0 J
10. A net force of 5.00 newtons is used to displace a 3.00 kg box a distance of 4.00 m in 2.00 second. How much work is done on the box?
    a) 6.00 J b) 8.00 J c) 15.0 J d) 20.0 J
11. A net force of 6.00 N is used to move a box 5.00 m in 2.00 s. How much power is used?
    a) 1.20 W b) 12.0 W c) 13.0 W d) 15.0 W
12. The motor of a snowmobile has an output of 1.00 kW. If the motor applies a force of 650 N to move the snowmobile at constant speed, what is the magnitude of the velocity of the snowmobile?
    a) 1.5 m/s b) 650 m/s c) 6.5×10^5 m/s d) 3.0 m/s
13. Two skaters with masses of 55.0 kg and 85.0 kg, respectively, face each other on the ice. They are initially at rest, but they push off each other in such a way that they move away from each other in opposite directions. If the speed of the lighter skater after they push off is 3.50 m/s, what is the speed of the 85.0 kg skater?
    a) 5.41 m/s b) 2.26 m/s c) 8.57 m/s d) 1.21 m/s
14. A worker exerts 45 W of power while lifting a 10.0 N box to a height of 1.0 m. How long does this take?
    a) 35 s b) 4.5 s c) 0.22 s d) 0.02 s

Part III—Calculations

Perform each of the following calculations.

15. At what height would a 5.5 kg bowling ball have 30.0 J of gravitational kinetic energy? (Assume g = 9.81 m/s^2.)
16. At what speed (the magnitude of the velocity) would a truck with a mass of 1.1×10^4 kg have a kinetic energy of 5.0×10^4 J?
17. A compressed spring with a constant of 38 N/m has 5.90 J of elastic potential energy. How far has it been compressed from its equilibrium position?
18. A girl does 48 J of work pushing her bicycle with a steady force of 15.0 N in the direction of its motion. How far does the bicycle move during this time?
19. Calculate the magnitude of the momentum of a football player with a mass of 140 kg and a velocity of 5.0 m/s.
20. A tennis player exerts a force of 35 N for a period of 0.15 s on a tennis ball with a mass of 0.10 kg. Calculate the change in the velocity of the tennis ball.

Answer Key

The actual answers will be shown in brackets, followed by an explanation. If you don't understand an explanation that is given in this section, you may want to go back and review the lesson that the question came from.

Lesson 3–1 Review

1. [work]—Remember, only the force along the axis of the displacement is part of the work.
2. [138 J]—$W = Fd\cos\theta = (20.0\text{ N})(12.0\text{ m})(\cos 55.0^\circ) = 138\text{ J}$
3. [2.38 m]—$d = \dfrac{W}{mg\cos\theta} = \dfrac{350\text{ J}}{(15.0\text{ kg})(9.81\text{ m/s}^2)(\cos 0^\circ)} = 2.38\text{ m}$

Lesson 3–2 Review

1. [kinetic energy]—If an object is not in motion, its velocity is zero, and so is the kinetic energy.

2. [9.5 J]—$P.E._{elas} = \frac{1}{2}kx^2 = \left(\frac{1}{2}\right)(25\ N/m)(0.874\ m)^2 = 9.5\ J$

3. [7.35×10^5 J]— $K.E. = \frac{1}{2}mv^2 = (0.5)(1.20 \times 10^3\ kg)(35.0\ m/s)^2 = 7.35 \times 10^5\ J$

Lesson 3–3 Review

1. [353 J]—Because the ramp is frictionless, the law of conservation of mechanical energy tells us that the kinetic energy of the box at the bottom of the ramp will be equal to the gravitational potential energy that it had at the top of the ramp.

   $K.E. = P.E._{grav} = mgh = (12.0\ kg)(9.81\ m/s2)(3.00\ m) = 353\ J$

2. [7.35 m/s]—We set the formulas for kinetic energy and gravitational potential energy equal to each other,

   $mgh = \frac{1}{2}mv^2$,

   and then solve for velocity to get the working formula $v = \sqrt{2gh}$.

   $v = \sqrt{2gh} = \sqrt{2(9.81\ m/s^2)(2.75\ m)} = 7.35\ m/s$

   Notice that the mass of the apple plays no part in the equation. Despite what Aristotle thought, heavy objects do *not* fall faster than light ones.

3. [49.1 J]—At the top of the shelf, the bowling ball's gravitational potential energy would be $P.E._g = mgh = (5.00\ kg)(9.81\ m/s^2)(2.00\ m) = 98.1\ J$, and the kinetic energy would be zero. When it has fallen 1.00 m, halfway to the floor, half of its gravitational potential energy will have turned into kinetic energy (98.1 J/2) = 49.1 J.

Lesson 3–4 Review

1. [power]—$\text{power} = \frac{\text{work}}{\text{change in time}}$

2. [21.6 s]—$\Delta t = \frac{mgh}{P} = \frac{(355\ kg)(9.81\ m/s^2)(15.5\ m)}{2.50 \times 10^3\ W} = 21.6\ s$

3. [240 W]—$P = \frac{F_w d}{\Delta t} = \frac{(340\ N)(3.50\ m)}{5.0\ s} = 240\ W$

Lesson 3–5 Review

1. [elastic collision]—Think of elastic collision as the type of collision where the objects bounce apart. When they stick together, that is an inelastic collision.
2. [momentum]—$p = mv$
3. [35 m/s]—$v = \frac{p}{m} = \frac{3.50\ \text{kg} \cdot \text{m/s}}{0.10\ \text{kg}} = 35\ \text{m/s}$

Lesson 3–6 Review

1. [impulse]—$J = F\Delta t$
2. [3.00×10^2 N · s]—$J = F\Delta t = (20.0\ \text{N})(15.0\ \text{s}) = 3.00 \times 10^2\ \text{N} \cdot \text{s}$
3. [44.8 m/s]—$v_f = v_i + \frac{F\Delta t}{m} = 3.50\ \text{m/s} + \frac{(12.0\ \text{N})(0.344\ \text{s})}{0.100\ \text{kg}} = 44.8\ \text{m/s}$

Chapter 3 Examination

1. [g. power]—Remember, the formula for power is $\text{power} = \frac{\text{work}}{\text{change in time}}$.
2. [f. mechanical energy]—Remember the law of conservation of mechanical energy.
3. [h. momentum]—Momentum is calculated with the formula: $p = mv$.
4. [c. kinetic energy]—If an object's velocity is zero, its kinetic energy is also zero.
5. [i. impulse]—$J = F\Delta t$
6. [b. watts]—Power is measured in watts (1 watt = 1 joule/second).
7. [d. quadrupled]—Remember, the kinetic energy of an object is proportional to the square of its velocity. $\text{K.E.} = \frac{1}{2}mv^2$
8. [a. 0.127 kg]—We know that $\text{P.E.}_g = mgh$, so

   $$m = \frac{\text{P.E.g}}{gh} = \frac{0.500\ \text{J}}{(9.81\ \text{m/s}^2)(0.400\ \text{m})} = 0.127\ \text{kg}$$

9. [b. 0.400 J]—When the apple has fallen half the distance, the value for height (h) in the equation $\text{P.E.}_g = mgh$ will be half the original value.

10. [d. 20.0 J]—Questions such as number 10 are designed to throw off people who simply take whatever numbers there are in the question and pick an operation to do at random. The only quantities we need to solve this one are the force and displacement. $W = Fd = (5.00\text{ N})(4.00\text{ m}) = 20.0\text{ J}$

11. [d. 15.0 W]—$P = \dfrac{W}{\Delta t} = \dfrac{Fd}{\Delta t} = \dfrac{(5.00\text{ N})(6.00\text{ m})}{2.00\text{ s}} = 15.0\text{ W}$

12. [a. 1.5 m/s]—Convert: $1.00\text{ kW} = \dfrac{1000\text{ W}}{1\text{ kW}} = 1.00 \times 10^3\text{ W}$

Remember, another formula for power is P = fv, so

$v = \dfrac{P}{f} = \dfrac{1.00 \times 10^3\text{ W}}{650\text{ N}} = 1.5\text{ m/s}$.

13 [b. 2.26 m/s]—$v_{2f} = -\left(\dfrac{m_1 v_{1f}}{m_2}\right) = -\left(\dfrac{55.0\text{ kg} \cdot 3.50\text{ m/s}}{85.0\text{ kg}}\right) = -2.26\text{ m/s}$. Because the question only asks for the speed, we will only report the magnitude of the velocity, 2.26 m/s.

14. [c. 0.22 s]—$\Delta t = \dfrac{W}{P} = \dfrac{F_w h}{P} = \dfrac{(10.0\text{ N})(1.0\text{ m})}{45\text{ W}} = 0.22\text{ s}$

15. [0.56 m]—$h = \dfrac{P.E._g}{mg} = \dfrac{30.0\text{ J}}{(5.5\text{ kg})(9.81\text{ m/s}^2)} = 0.56\text{ m}$

16. [3.0 m/s]—$v = \sqrt{\dfrac{2K.E.}{m}} = \sqrt{\dfrac{2(5.0 \times 10^4\text{ J})}{1.1 \times 10^4\text{ kg}}} = 3.0\text{ m/s}$

17. [0.56 m]—$x = \sqrt{\dfrac{2P.E._{elas}}{k}} = \sqrt{\dfrac{2(5.90\text{ J})}{38\text{ N/m}}} = 0.56\text{ m}$

18. [3.2 m]—$d = \dfrac{W}{F} = \dfrac{48\text{ J}}{15\text{ N}} = 3.2\text{ m}$

19. [$7.0 \times 10^2\text{ kg} \cdot \text{m/s}$]—$p = mv = (140\text{ kg})(5.0\text{ m/s}) = 7.0 \times 10^2\text{ kg} \cdot \text{m/s}$

20. [53 m/s]—$\Delta v = \dfrac{F\Delta t}{m} = \dfrac{(35\text{ N})(0.15\text{ s})}{0.10\text{ kg}} = 53\text{ m/s}$

Rotational and Circular Motion

Do you realize that you probably conducted physics experiments in the park before you were old enough to go to school? Your experiences with seesaws, merry-go-rounds, and swings offered early opportunities to learn the principles of physics involved in rotational and circular motion. Years later, even if you no longer visit parks, you are still surrounded by examples of objects that exhibit these principles. From swinging doors to spinning CDs, we are surrounded by examples of rotational and circular motion. In this chapter we will explore the terms and formulas related to these types of motion.

Lesson 4–1: Rotational Motion

Rotational motion is the motion of an object around an axis. This differs from translational motion, which we have been studying up until now, in that a rotating object may be in motion around an axis without showing an overall displacement with reference to a nearby object. A hockey puck sliding across a patch of ice shows translational motion. If the puck is also spinning, it has rotational motion as well.

A CD spinning in a stationary CD player shows rotational motion, but not translational motion, at least with reference to other objects in the room.

Many children conduct informal experiments involving rotational and circular motion at the park while riding on the merry-go-round. Whether riding on the type of merry-go-round found at many parks, which you have to push, or the carousel with horses found at amusement parks, you might recall that you seemed to travel faster when you sat near the outer rim of the circle. Just a few days before writing this lesson, I saw my son and daughter riding a merry-go-round at the park. My daughter, who is younger, sat near the center and had no trouble staying on. My son sat near the outer edge, where he found it difficult to stay on. He needed to hold on tight to avoid being hurled from the ride because he was actually moving faster than his sister, even though she was on the same ride!

Looking back at your own experience, does it seem odd to you that you can have two people on the same merry-go-round, which is rotating at a certain speed, and yet the two people can be traveling at different speeds? To explain this, let's start off by making sure that certain definitions are clear.

When we talk about the motion of the merry-go-round itself, we are talking about **rotational motion**. The merry-go-round experiences rotational motion, as it rotates about its axis of rotation, or pivot point. In this way, it is similar to the seesaw, only with a different orientation. The horizontal orientation of a seesaw only allows for a limited range of rotation, because one end or the other will eventually hit the ground. The vertical orientation allows the merry-go-round to rotate completely around or, as some might say, 360°.

You are probably accustomed to describing circles or angles in terms of degrees, but in this chapter we will be talking about another common unit called radians (rad). Just as a pizza can be cut into six, eight, or even 37 slices, a circle can be divided into any number of pieces. We usually break up a circle into 360 pieces, and call them degrees. We could also break the same circle into 6.28 pieces and call them radians.

Conversions

$$360^\circ = 6.28 \text{ rad}$$

$$1 \text{ rad} = 57.3^\circ \text{ (approximately)}$$

It may seem odd to break a circle up into an uneven number of pieces. Why are there 6.28 radians per circle? Did you notice that 6.28 is twice 3.14, the standard value of pi to three digits? Is there a connection between pi and radians beyond the fact that they are both concerned with circles?

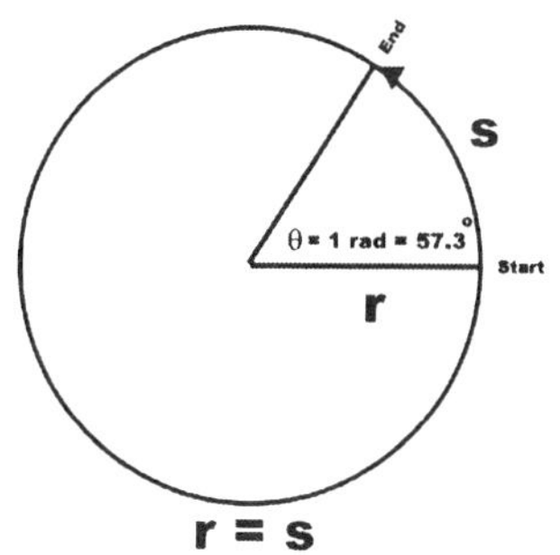

Figure 4.1

A radian is defined as the measure of an angle whose arc length (s) is equal to its radius (r). When a circular object, such as a merry-go-round, has rotated through an angle where the arc length of its path is equal to its radius, we say that it has rotated through an angle (θ) equal to one radian, or that it has an angular displacement of 1 rad.

We use radians to measure the angular displacement of an object that experiences rotational motion.

Angular Displacement

The change of position of a rotating body as measured by the angle through which it rotates.

$$\text{angular displacement } (\theta) = \frac{\text{change in arc length } (\Delta s)}{\text{radius } (r)}$$

or

$$\theta = \frac{\Delta s}{r}$$

Example 1

Find the angular displacement in radians of a bug sitting 3.5 cm from the center of a rotating CD as it traces out an arc length of 8.3 cm.

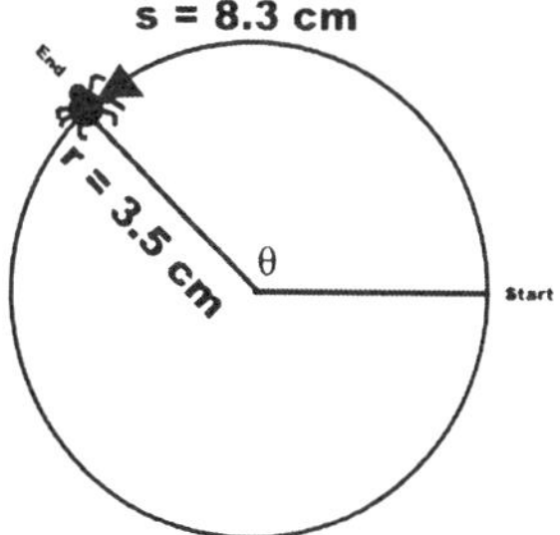

Figure 4.2

Given: r = 3.5 cm   Δs = 8.3 cm

Find: θ

Solution:

$$\theta = \frac{\Delta s}{r} = \frac{8.3 \not{cm}}{3.5 \not{cm}} = 2.3714285 = \mathbf{2.4\ rad}$$

Now, suppose a merry-go-round rotates completely around in what we might call a 360° circle. The arc length (s) it traces out would be equal to the diameter of the entire circle. The formula for the diameter of a circle is $2\pi r$. Let's see what happens when we put that value in for the arc length (s) in our angular displacement formula.

$$\theta = \frac{\Delta s}{r} = \frac{2\pi \not{r}}{\not{r}} = 2\pi$$

So, an angular displacement of 360° represents 2π rad, or 6.28 rad, and that is why there are 6.28 rad in every circle.

If you were asked to describe the speed of the merry-go-round, you would probably describe it in terms of its angular speed. The **angular velocity** of an object is the rate at which it rotates around its axis in a particular direction. This quantity, which is usually measured in radians per second (rad/s), can be determined by the following formula.

Angular Velocity

$$\text{angular velocity } (\omega) = \frac{\text{angular displacement } (\Delta\theta)}{\text{change in time } (\Delta t)}$$

or

$$\omega = \frac{\Delta\theta}{\Delta t}$$

Most sources will consider an object rotating clockwise to have a negative angular velocity. When the direction of the rotation is not considered, then the quantity may be referred to as **angular speed**.

Example 2

As a merry-go-round rotates in a clockwise direction around its axis, it undergoes an angular displacement of 9.80 radians in 2.0 seconds. What is the angular velocity of the merry-go-round?

Given: $\Delta\theta = -9.80$ rad $\quad \Delta t = 2.0$ s

Find: ω

Solution: $\omega = \frac{\Delta\theta}{\Delta t} = \frac{-9.80 \text{ rad}}{2.0 \text{ s}} = \mathbf{-4.9 \text{ rad/s}}$

The negative sign in our answer indicates that the angular velocity is in the clockwise direction. To help you visualize how fast this is, you could always convert radians to degrees

$$\left(\frac{4.9\ \not{rad}}{s} \times \frac{57.3°}{\not{rad}} = \mathbf{280°/s}\right)$$

and see that the merry-go-round is completing a little bit less than 80 percent of a rotation every second.

Two other terms that we should discuss at this point are **period** and **frequency**. Period (T) is the amount of time it takes to complete one cycle of motion. The term *period*, as it relates to rotational and/or circular motions, is the amount of time it takes for an object to complete a full rotation or revolution, usually measured in seconds. If it takes a merry-go-round 3.4 seconds to go around one time, it has a period of 3.4 s. Frequency is the inverse of period. Frequency (f) is the number of cycles of motion that take place each second. The term *frequency*, as it relates to rotational and/or circular motions, is the number of rotations or revolutions completed in one second. Frequency is sometimes described with the units cycles/second, rotations/second, or revolutions/second, but it is most commonly shown with the generic unit called hertz (Hz).

$$1\ \text{Hertz (Hz)} = \frac{1\ \text{cyle}}{\text{second}}$$

Example 3

A wagon wheel has an angular velocity of 7.50 rad/s. Calculate the period and frequency of the wheel.

Given: $\omega = 7.50$ rad/s $\quad \Delta\theta = 1$ cycle $= 6.28$ rad

Find: T and f

Solution: $T = \Delta t = \frac{\Delta\theta}{\omega} = \frac{6.28\ \not{rad}}{7.50\ \not{rad}/s} = \mathbf{0.837\ s}$

$$f = \frac{1}{T} = \frac{1}{0.837\ s} = 1.19\ s^{-1} = 1.19\ \text{cycles/s} = \mathbf{1.19\ Hz}$$

Look at the answers to Example 3 and see if you understand them. The wheel has an angular velocity of 7.50 rad/s. A complete rotation represents 6.28 rad, so this wheel is rotating around completely (period) every 0.837 s. It completes almost 1.2 rotations (frequency) every second. Notice that you could have solved for frequency first by dividing the angular velocity by the number of radians in one cycle or rotation.

$$f = \frac{\omega}{\Delta\theta} = \frac{7.50\ \cancel{rad}/s}{6.28\ \cancel{rad}} = 1.19\ s^{-1} = \mathbf{1.19\ Hz}$$

Then, we would find the period by taking the inverse of the frequency.

$$T = \frac{1}{f} = \frac{1}{1.19\ s^{-1}} = \mathbf{0.837\ s}$$

Now try the following review problems.

Lesson 4–1 Review

1. ______________ is the rate at which a body rotates in a particular direction.
2. Convert 5.81 rad to degrees.
3. Find the angular velocity of a bicycle tire that undergoes an angular displacement of 55.7 rad in 11.5 s.

Lesson 4–2: Torque

What causes a rotational motion, as described in our last lesson? Newton's first law reminds us that an object at rest remains at rest, and an object in motion remains in motion, unless acted upon by an unbalanced force. Applying this law to rotational motion we come to realize that if an object is rotating, no external force is required to keep it rotating. However, if a rotating object was initially at rest, an unbalanced force must have been involved at some point to get it to start rotating. Keep in mind, not every force will cause a rotational motion. It is possible for you to throw a football without spinning it, causing only translational motion. Or you can throw a football with what they call a "spiral," giving it both translational and rotational motion. In both cases, you applied an unbalanced force, but you only would have applied **torque** in the later case.

Torque (τ)

A measure of the ability of a force to rotate an object around an axis.

Torque = distance of force from axis × force × sine of angle between the force and lever

$$\tau = dF\sin\theta$$

Do you remember playing on a seesaw when you were a child? Did you encounter situations where you were on one side of the seesaw and a child with a much different weight was on the other? Can you recall what you had to do to compensate for the fact that you both were different weights? By trial and error, you may have found that you could compensate for the different weights by having the lighter child move further back on one side of the seesaw, away from the center, and having the heavier child move up, closer to the center. By shifting your positions on the seesaw, you may have been able to achieve **rotational equilibrium**, allowing you to keep the seesaw horizontal and motionless. You did this by making sure that the net **torque** on the seesaw was zero.

Torque is measured in N · m. If the net torque on an object such as a seesaw is zero, it will keep on doing what it is doing (maintain constant angular velocity), and the object is said to be at rotational equilibrium. When the net torque on an object that is free to rotate around an axis or pivot point is not equal to zero, there will be an angular acceleration. It is the net torque that we apply that allows us to open doors and turn screws and bolts. It was the net torque applied by your weight that allowed the seesaw to move down when you played at the park.

Figure 4.3 illustrates how to measure the distance between the pivot point (axis of rotation) and the applied force.

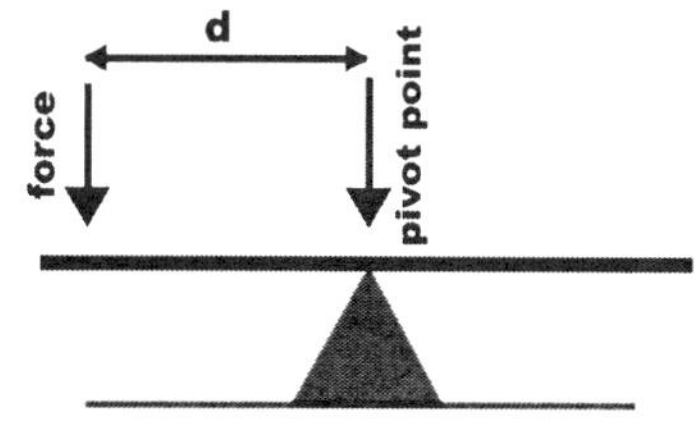

Figure 4.3

The reason why you needed the lighter child to move towards the end of the seesaw, away from the pivot point, is to maximize the torque on that side of the seesaw by increasing the value of d. Moving the heavier child inwards, towards the pivot point, decreases the torque on that side of the seesaw by decreasing the value of d.

At rotational equilibrium, $d_1F_1\sin\theta_1 = d_2F_2\sin\theta_2$

There are many similar real-life examples that involve torque. For example, if you want to open a very heavy swinging door, you should apply the force as far away from the pivot point (hinges) as possible. If you want to lift a heavy object with a lever, you want to apply the force as far away from the pivot point (fulcrum) as you can. If you are trying to remove a stubborn lug nut from your tire, you want to apply the force at the end of the handle of the tire iron. In each of these cases you may also maximize the torque by applying the force perpendicular to the surface or lever arm.

Example 1

A woman applies a force of 185 N perpendicular to the handle of a wrench at a distance of 85 cm from center of the nut (axis of rotation) she is turning. Calculate the magnitude of the applied torque.

Convert: $85\,\cancel{\text{cm}} \times \dfrac{1\text{ m}}{100\,\cancel{\text{cm}}} = \mathbf{0.85\text{ m}}$

Given: F = 185 N $\theta = 90°$ d = 0.85 m

Find: τ

Solution:

$$\tau = Fd\sin\theta = (185\text{ N})(0.85\text{ m})(\sin 90°)$$
$$= 157.25\text{ N}\cdot\text{m} = \mathbf{160\text{ N}\cdot\text{m}}$$

Suppose the same woman applied a force with the same magnitude on the same wrench at the same distance from the axis of rotation, but she didn't apply the force perpendicular to the (lever arm) wrench? How would that change our answer?

Example 2

A woman applies a force of 185 N at an angle of 60.0° to the handle of a wrench at a distance of 85 cm from center of the nut (axis of rotation) she is turning. Calculate the magnitude of the applied torque.

Convert: $85\,\cancel{\text{cm}} \times \dfrac{1\text{ m}}{100\,\cancel{\text{cm}}} = \mathbf{0.85\text{ m}}$

Given: F = 185 N $\theta = 60°$ d = 0.85 m

Find: τ

Solution:
$$\tau = \text{Fd}\sin\theta = (185\text{ N})(0.85\text{ m})(\sin 60^\circ)$$
$$= 136.18\text{ N}\cdot\text{m} = \mathbf{140\text{ N}\cdot\text{m}}$$

So, the woman would be exerting the same effort, but only some of her effort would translate to torque. Where would the rest of her effort be directed? Into trying to push the nut away from her, parallel to the direction of the handle.

Torque problems become more complex as multiple forces are applied. Sometimes multiple forces work together, and sometimes they oppose each other. To account for this, torques that would cause clockwise rotation are usually considered negative, while torques that would cause counterclockwise rotation are considered positive.

Example 3

A child with a mass of 35.0 kg sits on the right side of a horizontal seesaw, at a distance of 1.2 m from the pivot point. How far to the left of the pivot point should a 25.0 kg child sit in order to achieve rotational equilibrium?

Convert:

We will use Newton's second law and the mass of the children, which will be equal to the forces that they will exert on the seesaw. Because the seesaw is horizontal, the force of each weight will be applied perpendicular to it.

$F_1 = Fw_1 = mg = (35.0\text{ kg})(9.81\text{ m/s}^2) = 343.35\text{ N} = \mathbf{343\text{ N}}$

$F_2 = Fw_2 = mg = (25.0\text{ kg})(9.81\text{ m/s}^2) = 245.25\text{ N} = \mathbf{245\text{ N}}$

Given: $\tau_{net} = 0\text{ N}\cdot\text{m}$ $F_1 = 343\text{ N}$ $d_1 = 1.2\text{ m}$ $\theta_1 = 90^\circ$
$F_2 = 245\text{ N}$ $\theta_2 = 90^\circ$

Find: d_2

Isolate:

We start with rotational equilibrium: $\tau_{net} = \tau_1 + \tau_2 = 0\text{ N}\cdot\text{m}$

Subtracting τ_1 from both sides, we get: $\tau_2 = -\tau_1$

Which is equivalent to: $F_2 d_2 \sin\theta_2 = -(F_1 d_1 \sin\theta_1)$

Dividing both sides by $F_2 \sin\theta_2$ gives us: $d_2 = \frac{-(F_1 d_1 \sin\theta_1)}{F_2 \sin\theta_2}$

Solution: $d_2 = \frac{F_1 d_1 \sin\theta_1}{F_2 \sin\theta_2} = \frac{-(343\text{ N})(1.20\text{ m})(\sin 90^\circ)}{(245\text{ N})(\sin 90^\circ)} = \mathbf{-1.68\ m}$

So, the second child should sit 1.68 m to the left of the pivot point. The negative sign in our answer indicates that this child should be on the opposite side of the pivot point as the first child.

Lesson 4–2 Review

1. ______________ is the ability of an applied force to produce rotational motion.
2. How far from a pivot point should a 25.0 N force be applied perpendicularly to the surface of a lever to achieve a torque of 17.5 N · m?
3. A force of 18.5 N is applied at 35.0° to the surface of a seesaw at a distance of 0.875 m from the pivot point. Find the torque produced by this force.

Lesson 4–3: Circular Motion

In Lesson 4–1 we talked about the example of children riding on a merry-go-round. We described the motion of the merry-go-round in terms of rotational motion, as it rotated around its axis. The children riding on the merry-go-round experience **circular motion**, as they trace out a circular path around the center of the merry-go-round.

If you were asked to describe the angular speed of the children as demonstrated in Lesson 4–1, you would find that both children would have the same angular speed, regardless of where they sat on the merry-go-round, because they complete one 360° (or 6.28 rad) circle in the same amount of time. Experience tells us that our apparent speed does depend on where we sit on these types of rides. When you ride further out from the center of the circle, you have a greater **tangential speed**. The tangential speed of an object is its instantaneous linear speed along the tangent to its circular path.

If a line were drawn from each child to the center of the merry-go-round, you would get the **radius** of the circular path in which each child travels. We can mathematically determine the tangential speed of two different children sitting on the merry-go-round and show that they can be traveling at different speeds, despite being on the same ride.

In order to explain why the two children on the same ride can have different tangential speeds, let us recall that the formula for linear speed is

$$\text{speed} = \frac{\text{displacement}}{\text{time}}.$$

It takes each child the same amount of time to complete one circular path. This amount of time is referred to as the **period** (T) and it is measured in seconds. In that period, each child covers a different distance. The distance that each child covers is equal to the diameter of the circle ($2\pi r$) that he or she travels around. And because the radius measured from each child to the center of the merry-go-round is different, the children will travel different distances in a given amount of time. The tangential speed of each child will be equal to the rate at which the child travels around the circle.

$$v = \frac{d}{\Delta t} = \frac{2\pi r}{T}$$

Example 1

A boy and a girl sit on a merry-go-round with a period of 3.5 seconds. The boy and girl sit 0.75 m and 0.35 m from the center of the merry-go-round, respectively. Calculate the tangential speed of each child.

Given: T = 3.5 s $r_b = 0.75$ m $r_g = 0.35$ m
Find: v_b and v_g

Solution: $$v_b = \frac{2\pi r_b}{T} = \frac{2(3.14)(0.75\text{ m})}{3.5\text{ s}} = \mathbf{1.3\ m/s}$$

$$v_g = \frac{2\pi r_g}{T} = \frac{2(3.14)(0.35\text{ m})}{3.5\text{ s}} = \mathbf{0.63\ m/s}$$

As you can see from Example 1, two children sitting on the same ride can have different tangential speeds. The tangential speed (or velocity) of a child will be proportional to his or her distance from the center of rotation. The next time that you ride on a carousel, think about how your selection of a horse will influence your tangential speed.

There is a convenient formula showing the relationship between angular speed and tangential speed.

Tangential speed = distance from axis × angular speed

or

$$v = \mathbf{r}\omega$$

You can use this formula when you want to find the tangential speed of a passenger on a rotating object with a known angular speed. Or we could use the formula to find the angular speed of the merry-go-round described in Example 2 by isolating the angular speed.

Example 2

A girl sitting 0.35 m from the center of a spinning merry-go-round has a tangential speed of 0.63 m/s. Find the angular speed of the merry-go-round.

Given: r = 0.35m v = 0.63 m/s

Find: ω

Solution: $\omega = \frac{v}{r} = \frac{0.63\ \text{m/s}}{0.35\ \text{m}} = \mathbf{1.8\ rad/s}$

Lesson 4-3 Review

1. ________________ is the instantaneous linear speed of an object in circular motion along the tangent to its circular path.
2. A girl swings a yo-yo over her head in uniform horizontal circles with a tangential speed of 12.5 m/s. The radius of the circular path is 0.566 m. Find the period of the yo-yo.
3. Find the tangential speed of a girl sitting 0.994 m from the center of a merry-go-round with an angular speed of 4.50 rad/s.

Lesson 4-4: Centripetal Force and Centripetal Acceleration

One more time, let's draw upon a fairly common experience and refer back to the days when you rode the merry-go-round at the park. When you first got on the ride it was stationary. It had an angular velocity of zero, and you, once you got seated, had a tangential velocity of zero. Then someone, perhaps a parent or an older sibling, started rotating the merry-go-round. You shouted, "Faster!" and they increased their efforts, increasing the angular velocity of the ride, and, as a result, increased your tangential velocity. Let's suppose the person eventually got the merry-go-round rotating as fast as he or she could make it go, and maintained that angular velocity for a long period of time. Would your tangential velocity keep changing? Of course it would!

This goes back to the "trick" question we discussed in Lesson 1–2, when we described velocity as "speed in a particular direction." Velocity includes both speed and direction, and as you go around the merry-go-round, the direction of your tangential velocity keeps changing. This means that even if the speed, frequency, and period of the ride remain the same, your tangential velocity is constantly changing as you change direction. Notice how the direction of the velocity vector keeps changing in Figure 4.4.

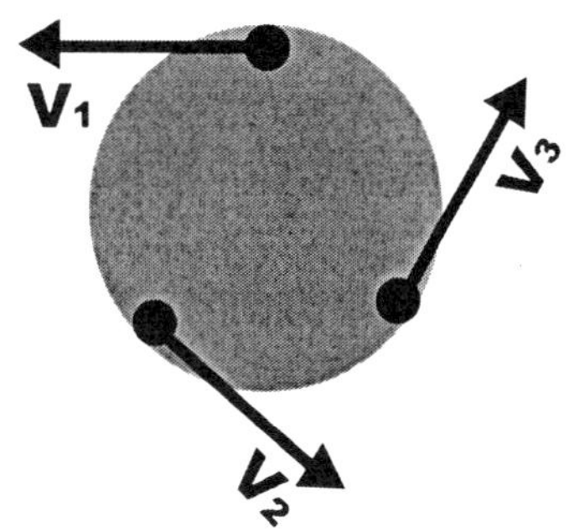

Figure 4.4

The child's velocity keeps changing, so the child, by definition, is accelerating.

$$\text{acceleration (a)} = \frac{\text{change in velocity } (\Delta v)}{\text{change in time } (\Delta t)}$$

Newton's second law tells us that this acceleration is evidence of an unbalanced force acting on the child.

Force = mass × acceleration

What is the source of this mysterious force and in what direction is it acting?

Can you imagine what would happen to a boy sitting on a merry-go-round if the friction between him and the ride was suddenly reduced to zero? The child's own inertia would cause him to fly off in the direction

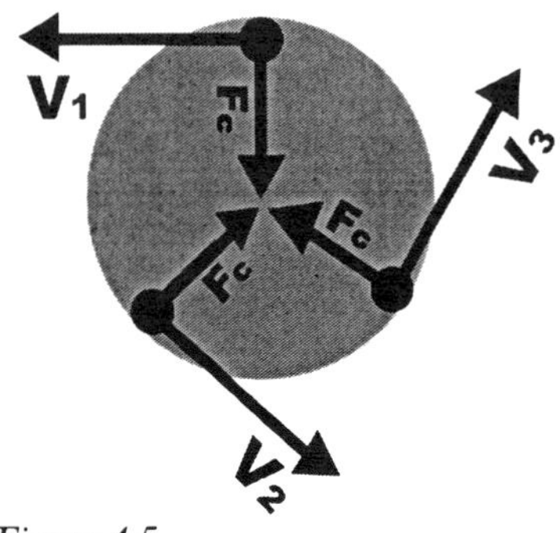

Figure 4.5

of his velocity at the instant that the ride became frictionless. Assuming that the boy is not holding on, it is only the force of friction between him and the ride that overcomes his inertia and changes his velocity. The force of friction and the acceleration experienced by the child are directed towards the center of rotation. We call this type of force and acceleration **centripetal**, or "center-seeking."

Centripetal Acceleration

An acceleration directed towards the center of a circular path.

$$a_c = \frac{v^2}{r} \text{ or } a_c = \frac{4\pi^2 r}{T}$$

Centripetal Force

A force directed towards the center of a circular path.

$$F_c = ma_c \text{ or } F_c = \frac{mv^2}{r} \text{ or } F_c = \frac{m4\pi^2 r}{T}$$

Example 1

A 35.0 kg boy sits on a merry-go-round with a period of 3.78 s at a distance of 175 cm from the center of rotation. Find the boy's tangential speed, and the centripetal acceleration and centripetal force he experiences.

Convert: $175\,\cancel{\text{cm}} \times \frac{1\text{ m}}{100\,\cancel{\text{cm}}} = \mathbf{1.75\ m}$

Given: m = 35.0 kg r = 1.75 m T = 3.78 s

Find: v, a_c, and F_c

Solution: $v = \frac{d}{\Delta t} = \frac{2\pi r}{T} = \frac{2(3.14)(1.75\text{ m})}{3.78\text{ s}} = \mathbf{2.91\ m/s}$

$$a_c = \frac{v^2}{r} = \frac{(2.91\ \text{m/s})^2}{1.75\ \text{m}} = \mathbf{4.84\ m/s^2}$$

$$F_c = ma_c = (35.0\ \text{kg})(4.84\ \text{m/s}^2) = \mathbf{169\ N}$$

There will be times when you know the angular velocity of the object and you will need to calculate the centripetal acceleration and/or force the object experiences. In these cases you will want to recall one of the formulas from our previous lesson: $v = r\omega$.

$$a_c = \frac{v^2}{r} = \frac{(r\omega)^2}{r} = r\omega^2$$

Example 2

A boy spins a 355 g ball on the end of a string with a length of 45.7 cm in a uniform horizontal circle above his head. The angular speed of the ball is 8.76 rad/s. Find the centripetal acceleration and tension in the string.

When the problem asks for the tension in the string, it is really asking for the centripetal force on the ball. The string is what provides the centripetal force in this example.

Convert: $355\ \text{g} \times \frac{1\ \text{kg}}{1000\ \text{g}} = \mathbf{0.355\ kg}$

$$45.7\ \text{cm} \times \frac{1\ \text{m}}{100\ \text{cm}} = \mathbf{0.457\ m}$$

Given: $m = 0.355$ kg $r = 0.457$ m $\omega = 8.76$ rad/s

Find: a_c and F_c

Solution:

$$a_c = \frac{v^2}{r} = \frac{(r\omega)^2}{r} = r\omega^2 = (0.457\ \text{m})(8.76\ \text{rad/s})^2$$
$$= \mathbf{35.1\ m/s^2}$$

$$F_c = ma_c = (0.355\ \text{kg})(35.1\ \text{m/s}^2) = \mathbf{12.4\ N}$$

Lesson 4–4 Review

1. ________________ is a force directed towards the center of a circular path.
2. Calculate the centripetal force that keeps a 65.0 kg object in a circular path with a centripetal acceleration of 23.5 m/s^2.
3. At what radius would an object in circular motion with a centripetal acceleration of 5.32 m/s^2 have a tangential speed of 22.4 m/s?

Lesson 4–5: Newton's Law of Gravity

I have heard it said that "Newton discovered gravity," which seems to suggest that there was no gravity before him. What Newton really did was show that it was the force of gravity that keeps the moon in orbit around Earth, and all of the planets in orbit around the sun. He went further to state that every object in the universe exerts a gravitational force on every other object in the universe. So all things, including you, can be thought to have a gravitational field around it. The force of gravitational attraction between two objects is proportional to the product of the masses of the objects, and inversely proportional to the square of the distance between them (measured from center of mass to center of mass). A proportionality constant (G), called the universal gravitational constant, has been added to the equation to allow for accurate calculations. The current value for G is approximately $6.67 \times 10^{-11}\ N \cdot m^2/kg^2$.

Universal Gravitational Constant

$$F_g = G\frac{m_1 m_2}{d^2}$$

G = universal gravitational constant = $6.67 \times 10^{-11}\ N \cdot m^2/kg^2$

One of the important things to keep in mind when studying the universal law of gravitation is that Newton's other laws are still in effect. We study these laws in isolation, but they exist all together. This means that Newton's third law, the law of action-reaction, plays an important part in the gravitational attraction between two bodies. If object A exerts a force of gravitational attraction on object B, object B exerts an equal and opposite force of gravitational attraction on object A. As we mentioned in Chapter 2, as hard as Earth is pulling you down, you are pulling the earth up. As the earth is pulling the moon towards it, the moon is pulling the earth with an equal and opposite force, causing tides.

Gravity is an example of a **field force**, or a force that acts over a distance. As you know, gravity can act over enormous distances, even allowing the sun to keep Earth in orbit. It is often useful to imagine the gravitational field around an object. As an object moves "deeper" into another gravitational field, it experiences a greater force. This explains why we weigh slightly more on the surface of Earth than we would on an airplane.

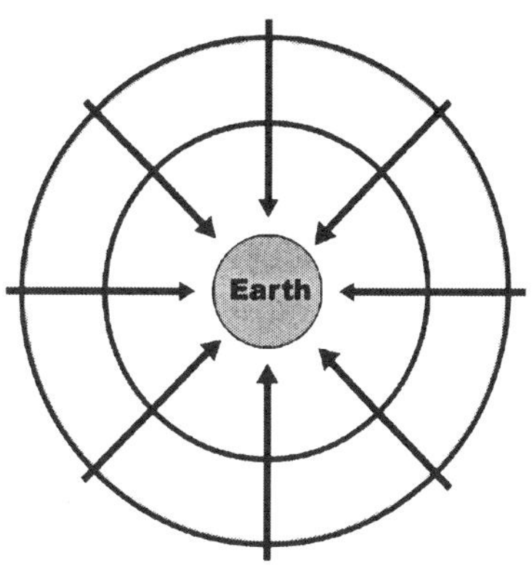

Figure 4.6

In a field diagram such as this one, the arrows indicate the direction of the force that an object will experience inside the field. The closer together the lines are at a given point, the greater the field strength at that point. As you can see, the lines get farther apart and the field gets weaker as you move away from the planet.

Newton's universal law of gravitation equation can be used to find the mass of unknown bodies, or the acceleration due to gravity (g) on other planets. This leads to many interesting problems in this area of study.

Example 1

Calculate the gravitational force of attraction between a 35.0 kg girl and a 1.20 × 10^3 kg car when they are separated by a distance of 3.45 m.

Given: $m_1 = 35.0$ kg $\quad m_2 = 1.20 \times 10^3$ kg $\quad d = 3.45$ m
$G = 6.67 \times 10^{-11}\ \mathrm{N \cdot m^2/kg^2}$

Find: F_g

Solution:

$$F_g = G\frac{m_1 m_2}{d^2} = \frac{(6.67 \times 10^{-11}\ \mathrm{N \cdot m^2/kg^2})(35.0\ \mathrm{kg})(1.20 \times 10^3\ \mathrm{kg})}{(3.45\ \mathrm{m})^2}$$

$$= \mathbf{2.35 \times 10^{-7}\ N}$$

Even if you were initially surprised to learn that all objects exert a gravitational force of attraction on all other objects, our answer for Example 1 should make you realize why we don't attract macroscopic objects in the same way that a very powerful magnet will draw nails or paper clips to it. The car in Example 1 may be considered a relatively massive object, and the distance between the girl and the car isn't far, but the attraction between the two due to gravity is only 2.35×10^{-7} N, or 0.000000235 N! Such a small force is not going to be enough to overcome the force of friction between the objects and the ground to draw them together.

You should always keep in mind that the gravitational force of attraction between two objects is inversely proportional to the *square* of the distance between their centers of mass. This means that if you halve the distance between two objects, you quadruple the force of gravity between them. If you triple the distance between the objects, you will decrease the force of gravity between them by a factor of ($3^2 = 9$) 9.

Example 2

Two objects separated by a distance of 4.0 cm exert a force of attraction due to gravity of 1.0 N on each other. What would be the gravitational force of attraction between these two objects if the distance between them was decreased to 1.0 cm?

Given: $d_i = 4.0$ cm $d_f = 1.0$ cm $F_i = 1.0$ N

Find: F_g

Solution: $F_g \propto \frac{1}{d^2} = \frac{1}{(1/4)^2} = \mathbf{16}$ $16 \times (1.0\ \text{N}) = \mathbf{16\ N}$

Example 3

The moon has a mass of approximately 7.349×10^{22} kg and a diameter of approximately 3476 kilometers. What would be the value of g near the surface of the moon?

Remember, the distance (d) that you need is measured to the center of the object, which would be the radius of the moon. The radius is half the diameter, so we find d:

$$r = \frac{\text{diameter}}{2} = \frac{3476\ \text{km}}{2} = 1738\ \text{km}.$$

Let's convert that to meters. $1738\ \cancel{km} \times \frac{1000\ m}{1\ \cancel{km}} = 1\,738\,000\ m$

Given: d = 1 738 000 m m = 7.349×10^{22} kg

G = 6.67×10^{-11} N · m²/kg²

Find: g_{moon}

Solution: $g_{moon} = G\frac{m}{d^2} = \frac{(6.67 \times 10^{-11}\ N \cdot m^2/kg^2)(7.349 \times 10^{22}\ kg)}{(1.738 \times 10^6\ m)^2}$

$= \mathbf{1.62\ m/s^2}$

You have probably heard that you would weigh 1/6th of your Earth weight on the moon, and now you have calculated it. Note,

$$\frac{9.81\ m/s^2}{1.62\ m/s^2} = 6.05.$$

Now try the review questions.

Lesson 4–5 Review

1. __________________ is a force that acts on objects without touching them.
2. An object has a weight of 200 N on the surface of Earth. How much would it weigh if its distance from the center of Earth were doubled?
3. Earth has a mass of approximately 5.97×10^{24} kg, and a radius of approximately 6.37×10^6 m. How much would a man who weighs 785 N on the surface of Earth weigh if he was flying in an airplane at an altitude of 1.00×10^4 meters? (Assume the plane is flying level, at a constant speed.)

Lesson 4–6: Kepler's Laws and the Motion of Satellites

When it comes to the motion of planets and other satellites, Johannes Kepler (1571–1630) wrote the rules. By assuming that the planetary orbits were elliptical, rather than circular, he was able to make data that had been collected earlier by Tycho Brahe (1546–1601) and other astronomers fit. You might wonder why we will discuss these elliptical orbits in a chapter about circular motion. One reason is because the gravitational force of attraction between the sun and a planet provides the centripetal force that keeps the planet in orbit, so that provides a connection to what we discussed in the previous lessons. Another reason is because if we treat the orbits of planets and other satellites as circular, we are able to come up with close approximations of their speeds and distances.

Kepler's First Law

The orbits of the planets are ellipses, with the sun at one focus of the ellipse.

Before Kepler, supporters of the Copernican heliocentric model of the solar system required a complex system of larger and smaller circles (orbits and epicycles) to account for the observed relative positions of the planets. From his observations of Mars, Kepler came to realize that the orbits of the planets were elliptical, allowing him to construct a model for the solar system that did not require epicycles.

Kepler's Second Law

The line joining the planet to the sun sweeps out equal areas in equal times as the planet travels around the ellipse.

As a planet moves about its elliptical orbit, with the sun at one focus, the distance between it and the sun is constantly changing. The position in the planetary orbit that brings the planet closest to the sun is called **perihelion**. The position farthest from the sun is called **aphelion**. In order for the planet to sweep out equal areas in equal times, it must move fastest at the perihelion and slowest at the aphelion.

Kepler's Third Law

The ratio of the squares of the revolutionary periods for two planets is equal to the ratio of the cubes of their semimajor axes.

$$\frac{T_1^2}{T_2^2} = \frac{a_1^3}{a_2^3}$$

The semimajor axis is half of the major axis, which, in turn, is the line that bisects the ellipse the long way. In other words, the **semimajor axis** is half of the longest axis of the ellipse.

Kepler's third law states that the ratio, $\frac{T^2}{a^3}$, is the same for every planet.

Chapter 4 Examination

Part I—Matching

Match the following terms to the definitions that follow.

a. radian	d. rotational motion	g. hertz
b. torque	e. centripetal force	h. frequency
c. period	f. centripetal acceleration	i. circular motion

_____1. The amount of time it takes to complete one vibration or one cycle of motion.

_____2. The motion of a body along a circular path.

_____3. A unit of angular measurement that is equal to 57.3°.

_____4. A unit that is equal to s^{-1}.

_____5. The ability of an applied force to produce rotational motion.

Part II—Multiple Choice

For each of the following questions, select the best answer.

6. Which of the following angles is equal to approximately 2.00 radians?

   a) 28.7° b) 57.3° c) 115° d) 360°

7. Convert 325° to radians.

   a) 5.67 rad b) 3.43 rad c) 1.11 rad d) 0.065 rad

8. The car of a Ferris wheel travels an arc length of 6.80 m. If the radius of the wheel is 5.25 m, what is the angular displacement of the car?

   a) 0.772 rad b) 1.30 rad c) 1.55 rad d) 12.1 rad

9. A bicycle tire is spinning with an angular velocity of 11.5 rad/s. How long would it take for a rock on the edge of the tire to experience an angular displacement of 4.50 rad?
   a) 0.391 s b) 2.56 s c) 7.00 s d) 51.8 s

10. A boy spins a yo-yo over his head with a uniform angular velocity of 7.85 rad/s. If the radius of the circular path is 0.350 m, find the tangential speed of the yo-yo.
    a) 22.4 m/s b) 8.20 m/s c) 7.50 m/s d) 2.75 m/s

11. A bug sitting 0.025 m from the center of a CD experiences a centripetal acceleration of 5.0 m/s^2. What is the bug's tangential speed?
    a) 0.354 m/s b) 0.125 m/s c) 200. m/s d) 5.025 m/s

12. A boy sitting 0.45 m from the center of a merry-go-round experiences a centripetal acceleration of 4.5 m/s^2. Find his angular speed.
    a) 5.0 rad/s b) 10.0 rad/s c) 3.2 rad/s d) 4.1 rad/s

13. Two objects separated by a distance of d experience a force of F due to gravity. If the distance between the objects is tripled, what would be the new gravitational force between them?
    a) 3F b) $\frac{F}{3}$ c) 9F d) $\frac{F}{9}$

14. Two objects separated by a distance of 0.025 m experience a force of 4.3×10^{-6} N due to gravity. If the mass of the first object is 35.0 kg, find the mass of the second object.
    a) 7.7 kg b) 1.2 kg c) 3.8 kg d) 4.4 kg

Part III—Calculations

Perform each of the following calculations.

15. If a girl sits 1.35 m from the center of a carousel with an angular speed of 3.89 rad/s, what is the centripetal acceleration she experiences?

16. A uniform meter stick is balanced on a pivot point at the 50.0 cm mark. A 1.0 N weight is hung from the 20.0 cm mark. Another weight will be hung at the 70.0 cm mark in order to establish rotational equilibrium when the meter stick is horizontal. What should the magnitude of this second weight be?

17. What is the period of a wave with a frequency of 5.0×10^{-3} Hz?
18. Calculate the centripetal force acting on a 0.445 kg toy being spun in uniform horizontal circles with a radius of 0.875 m and a period of 1.20 seconds.
19. Mars has a mass of approximately 6.421×10^{23} kg and a radius of approximately 3.397×10^{6} m. How much would a man with a mass of 80.0 kg weigh on Mars?
20. Calculate the gravitational force of attraction between a 1.5×10^{3} kg car and a 50.0 kg woman separated by a distance of 1.3 m.

Answer Key

The actual answers will be shown in brackets, followed by an explanation. If you don't understand an explanation that is given in this section, you may want to go back and review the lesson that the question came from.

Lesson 4–1 Review

1. [angular velocity]—The formula is:

$$\text{angular velocity } (\omega) = \frac{\text{angular displacement } (\Delta\theta)}{\text{change in time } (\Delta t)}$$

2. [333°]—$5.81 \cancel{\text{rad}} \times \frac{57.3^\circ}{1.00 \cancel{\text{rad}}} = 333^\circ$

3. [4.84 rad/s]—$\omega = \frac{\Delta\theta}{\Delta t} = \frac{55.7 \text{ rad}}{11.5 \text{ s}} = 4.84 \text{ rad/s}$

Lesson 4–2 Review

1. [torque]—The formula for torque is: $\tau = \text{dFsin } \theta$

2. [0.700 m]—$\text{d} = \frac{\tau}{\text{Fsin } \theta} = \frac{17.5 \cancel{\text{N}} \cdot \text{m}}{25.0 \cancel{\text{N}}} = 0.700 \text{ m}$

3. [9.28 N · m]—$\tau = \text{Fdsin } \theta = (18.5 \text{ N})(0.875 \text{ m})(\sin 35.0^\circ) = 9.28 \text{ N} \cdot \text{m}$

Lesson 4–3 Review

1. [tangential speed]

2. [0.285 s]—$T = \frac{2\pi r}{v} = \frac{2(3.14)(0.566\ \text{m})}{12.5\ \text{m/s}} = 0.285\ \text{s}$

3. [4.47 m/s]—$v = r\omega = (0.994\ \text{m})(4.50\ \text{rad/s}) = 4.47\ \text{m/s}$

Lesson 4–4 Review

1. [centripetal force]—The formula for centripetal force is $F_c = ma_c$.

2. [1530 N]—$F_c = ma_c = (65.0\ \text{kg})(23.5\ \text{m/s}^2) = 1530\ \text{N}$

3. [94.3 m]—$r = \frac{v^2}{a_c} = \frac{(22.4\ \text{m/s})^2}{5.32\ \text{m/s}^2} = 94.3\ \text{m}$

Lesson 4–5 Review

1. [field force]—Example of field forces are gravity, electrostatics, and magnetism.

2. [50 N]—Newton's law of universal gravitation tells us that the force of gravitational attraction between two objects due to gravity is inversely proportional to the distance between them. If you double the distance between the objects, the measurement of the weight will be divided by ($2^2 = 4$) 4.

   $$\frac{200\ \text{N}}{4} = 50\ \text{N}$$

3. [783 N]—Do you remember the difference between mass and weight? The weight of an object changes due to location, but the mass doesn't. Let's begin by finding the mass of the man, which will be the same at any altitude.

   $$m = \frac{F_w}{g} = \frac{785\ \text{N}}{9.81\ \text{m/s}^2} = 80.0\ \text{kg}$$

   Next, let's add the altitude of the aircraft to the radius of Earth to find the total distance between the man and the center of Earth.

   $6.37 \times 10^6\ \text{m} + 1.00 \times 10^4\ \text{m} = 6.38 \times 10^6\ \text{m}$

   Now we can use Newton's law of universal gravitation to calculate the weight of the man at this new distance.

   $$F_g = G\frac{m_1 m_2}{d^2} = \frac{(6.67 \times 10^{-11}\ \text{N}\cdot\text{m}^2/\text{kg}^2)(80.0\ \text{kg})(5.97 \times 10^{24}\ \text{kg})}{(6.38 \times 10^6\ \text{m})^2}$$

   $$= 783\ \text{N}$$

Chapter 4 Examination

1. [c. period]—If you get period mixed up with frequency, just think of "period of time."
2. [i. circular motion]
3. [a. radian]—One radian also equals the angle subtended at the center of a circle by an arc length equal to the length of the radius of the circle.
4. [g. hertz]—Frequency is measured in hertz.
5. [b. torque]—The torque is the product of the component of the force that causes rotation, and the length measured between the location of the force and the axis of rotation.
6. [c. 115°]—1 rad = 57.3°, so $2.00\,\cancel{\text{rad}} \times \dfrac{57.3^\circ}{1\,\cancel{\text{rad}}} = 115^\circ$
7. [a. 5.67 rad]—$325\cancel{^\circ} \times \dfrac{1\text{ rad}}{57.3\cancel{^\circ}} = 5.57\text{ rad}$
8. [b. 1.30 rad]—$\theta = \dfrac{\Delta s}{r} = \dfrac{6.80\,\cancel{\text{m}}}{5.25\,\cancel{\text{m}}} = 1.30\text{ rad}$
9. [a. 0.391 s]—$\Delta t = \dfrac{\Delta\theta}{\omega} = \dfrac{4.50\,\cancel{\text{rad}}}{11.5\,\cancel{\text{rad}}/\text{s}} = 0.391\text{ s}$
10. [d. 2.75 m/s]—$v = r\omega = (0.350\text{ m})(7.85\text{ rad/s}) = 2.75\text{ m/s}$
11. [a. 0.354 m/s]—$v = \sqrt{a_c r} = \sqrt{(5.0\text{ m/s}^2)(0.025\text{ m})} = 0.354\text{ m/s}$
12. [c. 3.2 rad/s]—$\omega = \sqrt{\dfrac{a_c}{r}} = \sqrt{\dfrac{4.5\,\cancel{\text{m}}/\text{s}^2}{0.45\,\cancel{\text{m}}}} = 3.2\text{ rad/s}$
13. [d. $\dfrac{F}{9}$]—From the formula $F_g = G\dfrac{m_1 m_2}{d^2}$, we see that the force of gravity decreases with the square of the distance. $3^2 = 9$, so the force is divided by 9.
14. [b. 1.2 kg]—Remember, $G = 6.67 \times 10^{-11}\text{ N}\cdot\text{m}^2/\text{kg}^2$. Isolating m_2 in our equation for universal gravitation, we get

$$m_2 = \frac{F_g d^2}{G m_1} = \frac{(4.3 \times 10^{-6}\text{ N})(0.025\text{ m})^2}{(6.67 \times 10^{-11}\text{ N}\cdot\text{m}^2/\text{kg}^2)(35.0\text{ kg})} = 1.2\text{ kg}.$$

15. [20.4 m/s²]—$a_c = r\omega^2 = (1.35\text{ m})(3.89\text{ rad/s})^2 = 20.4\text{ m/s}^2$

16. [1.5 N]—The key to this problem is to make sure you don't confuse position with distance. We need to find out how far the forces are going to be exerted from the pivot point.

$d_1 = 50.0\text{ cm} - 20.0\text{ cm} = 30.0\text{ cm}$

$d_2 = 50.0\text{ cm} - 70.0\text{ cm} = -20.0\text{ cm}$

$$F_2 = \frac{-(F_1 d_1 \sin\theta_1)}{d_2 \sin\theta_2} = \frac{-(1.0\text{ N})(30.0\text{ cm})(\sin 90^\circ)}{(-20.0\text{ cm})(\sin 90^\circ)} = 1.5\text{ N}$$

17. [2.00×10^2 s]—$T = \frac{1}{f} = \frac{1}{5.00 \times 10^{-3}\text{ Hz}} = 2.00 \times 10^2\text{ s}$

18. [12.8 N]—$F_c = ma_c = \frac{m4\pi^2 r}{T} = \frac{(0.445\text{ kg})(4\pi^2)(0.875\text{ m})}{1.20\text{ s}} = 12.8\text{ N}$

19. [297 N]—You need to calculate the value of the acceleration due to gravity (g_{mars}) on Mars, and multiply this value by the man's mass.

$$g_{mars} = G\frac{m}{d^2} = \frac{(6.67 \times 10^{-11}\text{ N}\cdot\text{m}^2/\text{kg}^2)(6.421 \times 10^{23}\text{ kg})}{(3.397 \times 10^6\text{ m})^2} = 3.71\text{ m/s}^2$$

$$Fw = mg = (80.0\text{ kg})(3.71\text{ m/s}^2) = 297\text{ N}$$

20. [3.0×10^{-6} N]—

$$F_g = G\frac{m_1 m_2}{d^2} = \frac{(6.67 \times 10^{-11}\text{ N}\cdot\text{m}^2/\text{kg}^2)(50.0\text{ kg})(1.50 \times 10^3\text{ kg})}{(1.3\text{ m})^2}$$

$$= 3.0 \times 10^{-6}\text{ N}$$

1 2 3 4 5 6 7 8 9 10

Electric Charges, Forces, and Fields

I remember going with my brother and sisters to visit my grandmother in her apartment that had a long, carpeted hallway. My siblings and I would spend much of our time shuffling up and down the hallway, building up static charges on our bodies, only to discharge these charges onto each other. A spark would be visible and a crackle would be audible as we shocked each of our victims. We never grew tired of the game, and to this day, I am still fascinated by the electric discharges that can be produced when an excess of electrons are built up on an object. In this chapter, we will explore some of the aspects of static electricity.

Lesson 5–1: Electric Charges

What people often refer to as **static electricity** or **static charge** is a buildup of charge on an object. How do objects obtain charges and what do people mean when they say that there are two *types* of charge?

If you ever played with bar magnets, then you know that you can put two like poles together, say positive to positive, and they will repel each other. You can also put two unlike poles together, one positive and one negative, and they will attract each other. Similarly, some charged objects repel each other and others attract each other. This evidence suggests that there are two different types of charges.

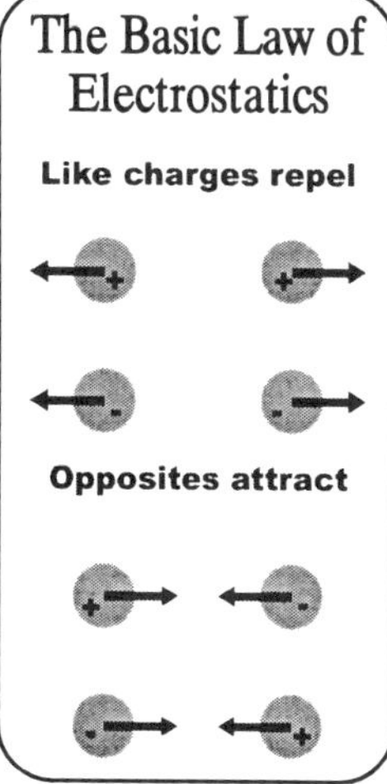

The Basic Law of Electrostatics
Like charges repel
Opposites attract

Figure 5.1

When you take clothes out of the drier and find a sock stuck to a sweater, we call this *static cling*. Static cling results when one item of clothing has a positive charge and the other has a negative charge. As is the case with magnets, unlike objects attract. But how does a sock obtain a charge?

Thinking back to your study of chemistry, you should recall two types of charged subatomic particles, the proton and the electron, with a positive charge and a negative charge, respectively. Atoms can gain or lose electrons to become **ions**, or charged atoms. A **negative ion** is an atom that has gained one or more extra electrons. A **positive ion** is an atom that has lost one or more electrons.

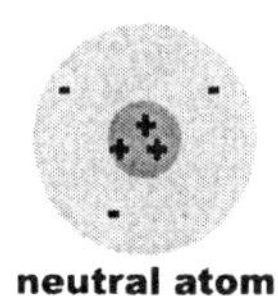

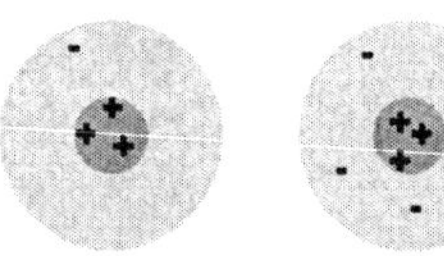

Figure 5.2

When we say that an atom, or any object, has *lost* an electron, we don't mean that it was destroyed. Electrons travel from atom to atom, or object to object, but they aren't created or destroyed in the process of normal charging. In most of the cases that we will discuss in this chapter, charging is accomplished by getting some electrons to leave one object and go to another, leaving the first object with a deficiency of electrons and a positive charge, and the second object with an excess of electrons and a negative charge.

The electron is thought to have the smallest amount of negative charge possible. The proton is thought to have the smallest possible amount of positive charge. These elementary charges (e) are equal in magnitude, but opposite in type. In physics we represent charge with the letter q. We measure charges in units called coulombs (C). One coulomb actually represents a relatively large charge, so the elementary charge, measured in coulombs, is incredibly small.

Elementary Charges

The charge on 1 electron (e^-) = -1.60×10^{-19} C

The charge on 1 proton (p^+) = $+1.60 \times 10^{-19}$ C

When you studied chemistry, you probably represented ions with superscripts, indicating the number of electrons they had gained or lost. For example, the sulfide ion (S^{2-}) has two extra electrons, giving it a net charge of –2. If you were asked to indicate the charge on this ion in coulombs, you would multiply the number of excess electrons (n) by the charge on one electron (e), as shown here:

$$q = n \times e = 2\cancel{e^-}(-1.60 \times 10^{-19}\ \text{C}/\cancel{e^-}) = -\mathbf{3.20 \times 10^{-19}\ C}.$$

If an ion is missing electrons, it has more protons then electrons, giving it a net positive charge. If you were asked to give the charge of a positive ion in coulombs, you multiply the number of excess protons by the charge on one proton.

Example 1

Determine the charge in coulombs of an ion of aluminum, Al^{3+}.

Given: Number of extra protons (n) = 3
Charge on one proton (e) = $+1.60 \times 10^{-19}$ C
Find: q

Solution: $q = n \times e = 3\cancel{P^+}(+1.60 \times 10^{-19}\ \text{C}/\cancel{P^+}) = +\mathbf{4.80 \times 10^{-19}\ C}$

As is the case with atoms, neutral macroscopic objects contain the same number of protons and electrons and have a net charge of zero. Charged objects have an unequal number of protons and electrons. If they have more electrons than protons, they have a net negative charge. If they have fewer electrons than protons, then they have a net positive charge.

Example 2

A woman walks across a carpet and obtains a charge of -4.56×10^{-16} C. How many excess electrons does this represent?

Given: $q = -4.56 \times 10^{-16}$ C $e = -1.60 \times 10^{-19}$ C

Find: number of e^-

Solution: $\text{number of } e^- = \dfrac{q}{e} = \dfrac{-4.56 \times 10^{-16}\ \cancel{C}}{-1.60 \times 10^{-19}\ \cancel{C}/e^-} = \mathbf{2850\ e^-}$

Remember, if the woman took 2850 electrons from the carpet, giving her a net charge of -4.56×10^{-16} C, the carpet would have a deficiency of 2850 electrons, giving it a net charge of $+4.56 \times 10^{-16}$ C, so the total charge would be conserved.

This brings us back to the sweater and sock example from the beginning of the lesson. As clothes come in contact with each other in the dryer, some items give up some of their electrons and obtain a net positive charge, while others take on some additional electrons, giving them a net negative charge. Consequently, oppositely charged items can stick together because of the electrostatic force of attraction between them.

Lesson 5–1 Review

1. ____________________ is a positively charged particle commonly found in the nucleus of an atom.

2. ____________________ is an atom that has gained one or more electrons.

3. An oxide ion (O^{2-}) has gained two additional electrons. What is the charge on such an ion, in coulombs?

Lesson 5–2: Electric Forces

Charged objects exert forces at a distance on other charged objects. If you rub an inflated rubber balloon on your hair, and then hold the balloon over some small bits of paper or salt crystals, you will see the balloon attract the objects from a distance. The paper bits or salt crystals will appear to leap off the table and stick to the balloon. This electrostatic force is another example of a field force. Similar to the force of gravity, it can act over a distance. Also, as with the case of gravitational attraction, electrostatic forces decrease with the square of the distance between the charged objects in question.

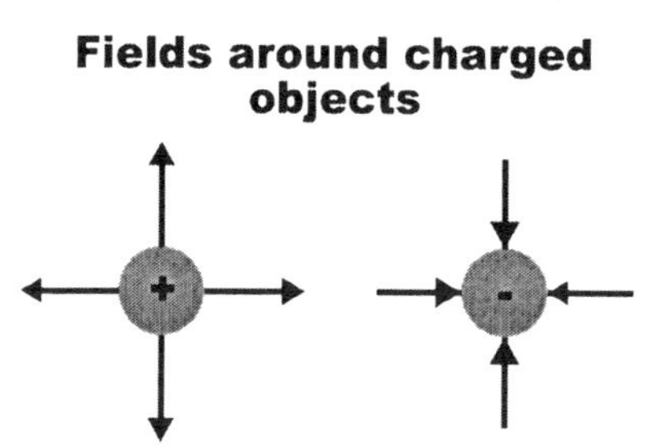

Figure 5.3

You will notice that the formula for **Coulomb's law**, which is used to calculate the force between charged objects, looks similar to Newton's universal law of gravitation. There are, however, some significant differences between the two laws, as summarized by the table on page 177.

Comparing and Contrasting Coulomb's Law and Newton's Universal Law of Gravitation

Formulas:	Coulomb's Law	Newton's Law
	$F = k_c \frac{q_1 q_2}{d^2}$	$F = G \frac{m_1 m_2}{d^2}$

Newton's law: The gravitational force of attraction between two objects is proportional to the product of the two masses and inversely proportional to the square of the distance between them.

Coulomb's law: The electrostatic force of attraction or repulsion between two charged objects is proportional to the product of the two charges and inversely proportional to the square of the distance between them.

Similarities:

- Both laws follow what is called the *inverse square law* for distance.
- Both formulas contain a proportionality constant.

Differences:

- The force calculated with Newton's law can only be positive, but the force calculated with Coulomb's law may be positive or negative.
- The proportionality constant used in Newton's law (G) is a very small number, 6.67×10^{-11} N · m²/kg², but the proportionality constant (k_c) used in Coulomb's law is a very large number, 9.0×10^{9} N · m²/C².

By comparing the proportionality constants required for the two laws in the table, you can gain an understanding of why gravity is considered a relatively weak force.

Example 1

Calculate both the electrostatic force (F_e) and the gravitational force (F_g) between a proton ($m_p = 1.67 \times 10^{-27}$ kg) and an electron ($m_e = 9.11 \times 10^{-31}$ kg) separated by a distance of 3.0×10^{-3} m.

Given: $q_1 = +1.60 \times 10^{-19}$ C $q_2 =$ “1.60×10^{-19} C
$m_1 = 1.67 \times 10^{-27}$ kg $m_2 = 9.11 \times 10^{-31}$ kg
$d = 3.0 \times 10^{-3}$ m $G = 6.67 \times 10^{-11}$ N · m²/kg²
$k_c = 9.0 \times 10^{9}$ N · m²/C²

Find: F_e and F_g

$$F_e = k_c \frac{q_1 q_2}{d^2}$$

Solution:

$$= \frac{(9.0 \times 10^{9}\ \text{N} \cdot \text{m}^2/\text{C}^2)(+1.60 \times 10^{-19}\ \text{C})(-1.60 \times 10^{-19}\ \text{C})}{(3.0 \times 10^{-3}\ \text{m})^2}$$

$$= \mathbf{-2.6 \times 10^{-23}\ N}$$

$$F_g = G \frac{m_1 m_2}{d^2}$$

$$= \frac{(6.67 \times 10^{-11}\ \text{N} \cdot \text{m}^2/\text{kg}^2)(1.67 \times 10^{-27}\ \text{kg})(9.11 \times 10^{-31}\ \text{kg})}{(3.0 \times 10^{-3}\ \text{m})^2}$$

$$= \mathbf{1.1 \times 10^{-62}\ N}$$

Note, the answer for the electrostatic force (F_e) is negative, indicating that unlike charges were involved, so the particles will attract each other. Also notice that the force of gravity (F_g) between the two particles is much less than the electrostatic force. How much stronger is the electrostatic force than the force of gravity in this example? Simply divide the smaller number into the larger:

$$\frac{(2.6 \times 10^{-23})}{(1.1 \times 10^{-62})} = 2.4 \times 10^{39}$$

So, the electrostatic force of attraction between the two particles in this problem is about 2 400 000 000 000 000 000 000 000 000 000 000 000 000 times as strong as the force of gravity between them! They aren't kidding when they say that gravity is a relatively weak force!

Even more interesting than the implications of the weakness of gravity are the implications of the strength of the electrostatic force. The bottom line is that the electrostatic force is so strong that we never actually touch anything, at least not in the way that most people think we do. Objects never really come in contact with each other. Matter is made up of atoms, and each atom is surrounded by an electron cloud. All electrons are like charges, which repel each other. The electrostatic force of repulsion between neutral atoms is so great that they can only come so close together without going through a chemical reaction and bonding. So, when we shake hands with someone, the atoms of our individual hands never touch. When we walk across the floor, the atoms of our shoes never actually touch the atoms of the floor. When we lie in bed, our atoms never actually touch the atoms of our sheets! You may want to think about this for a while before moving on to the next practice problem.

Example 2

A calcium ion (Ca^{2+}) and an aluminum ion (Al^{3+}) exert an electrostatic force of repulsion on each other with a magnitude of 3.0×10^{-28} N. How far apart are these ions?

Convert: First, let's calculate the charge on each ion. To do this, all we need to do is multiply the number of excess protons given by the superscript in each ion by the elementary charge (charge on each proton).

$$q_1 = (Ca^{2+}) = 2\,\cancel{P^+} \times (+1.60 \times 10^{-19}\ C/\cancel{P^+}) = \mathbf{+3.20 \times 10^{-19}\ C}$$

$$q_2 = (Al^{3+}) = 3\,\cancel{P^+} \times (+1.60 \times 10^{-19}\ C/\cancel{P^+}) = \mathbf{+4.80 \times 10^{-19}\ C}$$

Given: $q_1 = +3.20 \times 10^{-19}$ C $\quad q_2 = +4.80 \times 10^{-19}$ C

$F = 3.0 \times 10^{-28}$ N $\quad k_c = 9.0 \times 10^9\ N \cdot m^2/C^2$

Find: d

Isolate: $F = k_c \dfrac{q_1 q_2}{d^2}$

Multiply both sides by d^2: $d^2 \times F = k_c \dfrac{q_1 q_2}{\cancel{d^2}} \times \cancel{d^2}$

Divide both sides by F: $\dfrac{d^2 \times \cancel{F}}{\cancel{F}} = k_c \dfrac{q_1 q_2}{F}$

Take the square root of both sides: $\sqrt{d^2} = \sqrt{k_c \dfrac{q_1 q_2}{F}}$

We get: $d = \sqrt{k_c \frac{q_1 q_2}{F}}$

$$d = \sqrt{k_c \frac{q_1 q_2}{F}}$$

Solution: $$= \sqrt{\frac{(9.0 \times 10^9 \text{ N} \cdot \text{m}^2/\text{C}^2)(+4.80 \times 10^{-19} \text{ C})(+3.20 \times 10^{-19} \text{ C})}{3.0 \times 10^{-28} \text{ N}}}$$

$$= \mathbf{2.1\ m}$$

As a final point of interest, notice again that the electrostatic force between two charged objects is inversely proportional to the square of the distance between them. As the distance increases, the force between them decreases. While the force gets quite small as the distance increases, it doesn't approach zero until the distance between the objects approaches infinity. So objects don't really "escape" these fields, they just get to a point where the force exerted on them is not significant.

Lesson 5–2 Review

1. According to ________________, the electrostatic force between two charged objects is directly proportional to the product of their charges and inversely proportional to the square of the distance between them.
2. What would happen to the electrostatic force between two objects if the charge on each of them were doubled?
3. What would happen to the electrostatic force between two objects if the distance between them were doubled?

Lesson 5–3: Methods for Charging Objects

You have probably employed several methods to charge objects in the past. Did you ever rub a balloon on your head and then stick it to a wall? If so, you have performed two out of the three methods for charging objects that we will be learning in this lesson. Before we get to these methods, however, we will go over some preliminary information that relates to these methods of charging.

If you think of electrons as the carriers of flowing charge, you may realize that this charge will not flow with equal ease across all surfaces. Some materials, called **conductors**, allow electrons to flow freely through them. Metals are good conductors, which is why we use them for wiring and lightning rods. Other materials, called **insulators**, do not allow electrons to flow easily through them. Rubber and plastic are good insulators, and we use these materials to block the flow of electricity. For example, wires are covered with rubber insulation.

One important difference between insulators and conductors is what happens to excess charges on them. If you give a metallic object an excess negative charge, the electrons will spread out over the surface of the object, trying to maximize the distance between electrons. This occurs because the electrons are all like charges, so they repel each other. If the object is a uniform sphere, the charge spreads across the surface of the sphere until it is uniformly distributed. If the conductor doesn't have a uniform shape, charges tend to build up on the surface of points and protrusions.

For insulators, the charges aren't free to move. If you charge a particular area of the insulator, the charges can't migrate freely through or across it, so they tend to stay in that area until they are given an opportunity to leave. A balloon is made of rubber, a good insulator, so if you rub the balloon in your hair, it will strip excess electrons from your hair, leaving an area of the balloon with a net negative charge. These electrons will stay on that particular area unless a conductive path is provided to allow them to escape.

Dry air is a good insulator, but moist or humid air is a fairly good conductor. When charged objects are surrounded by dry air, they will hold on to their excess charge longer than if they were surrounded by humid air. One reason why static cling occurs in the dryer is because the air inside the dryer becomes (not surprisingly) very dry. Charged clothing items will retain a charge for longer and stick to another item.

People with straight hair suffer more "bad hair days" in dry weather, because their individual hairs will obtain like charges and repel each other. The result is what is sometimes called "fly-away" hair. If you are not familiar with this phenomenon, you may have seen it at the park instead. When my daughter slides down a plastic slide on a dry day, much of her hair starts to stand straight up. The reason for this effect can be understood better by looking at an instrument called an **electroscope**, which is used to determine the type of charge on an object.

The important parts of an electroscope to watch are the vanes. The vanes are usually made of some light metal foil. The metal foil is a good conductor, and it is attached to the knob. When a charged object comes near to, or touches, the knob, the vanes become charged. When the vanes are not charged, they hang together, but when they both obtain a like charge, they repel each other and hang more apart.

Discharged Electroscope

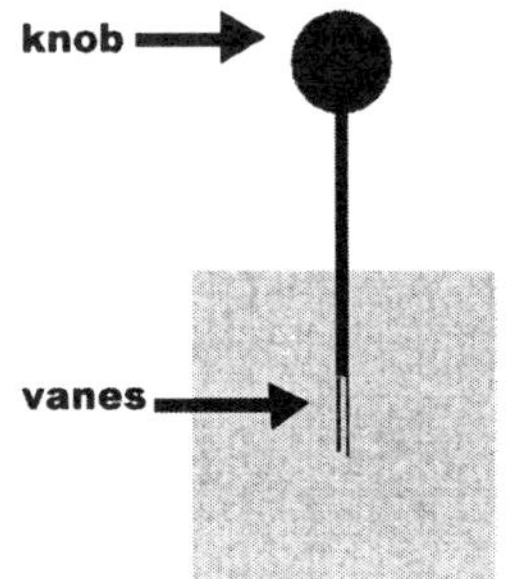

Figure 5.4

This is just what happens to some hair on dry days. The individual strands of hair obtain like charges and repel each other.

Charged Electroscope

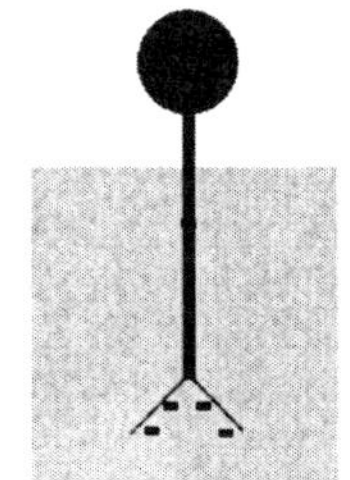

Figure 5.5

An electroscope can be used to determine the type of charge on an object when you don't know whether the object is positive or negative. Let's start off with the example of rubbing the balloon in your hair. I told you earlier in this lesson that this method will result in a balloon with an excess of electrons and a net negative charge on an area of the balloon's surface. If you bring this negatively charged area of the balloon close to a neutral electroscope, you will notice that the vanes of the electroscope move slightly apart, before the balloon even touches the knob. This happens through our first method of charging, called **polarization**.

Polarization is the process of inducing a temporary separation of charges in a neutral object by bringing it in close proximity to a charged object. Think of the knob and vanes of the electroscope. They are made of metal, and some of the electrons in metals are free to move. When you bring a negatively charged balloon near to the knob of the electroscope, the negative area of the balloon will repel the electrons in the knob. Because the electrons in the knob are free to move, some will migrate to the vanes in order to get away from the excess electrons on the balloon. As these electrons gather on the vanes, they repel each other, causing the vanes to move apart.

What makes this method of charging temporary is that if you move the balloon away from the electroscope, having never actually touched the knob with it, the excess electrons from the vanes will now redistribute themselves over the entire conductive surface, returning the electroscope to the uncharged state.

Now, let's go over the method of charging called **conduction**. Conduction is the process of charging a neutral object by bringing it in contact with a charged object. Let's suppose that we now take the area of the balloon with the excess electrons and actually touch the knob of the electroscope. Remember, the excess electrons on the balloon want to get away from each other. It is only the fact that the balloon is made of an insulator that prevents the electrons from spreading out. Now the electrons have come in contact with the conductive metal of the electroscope, and many of them will take this opportunity to spread out on to the surface of the electroscope. The vanes of the electroscope spread out again, indicating that they have obtained a charge. Now, even when we remove the balloon, the vanes of the electroscope stay open because the electroscope has a residual charge.

Before we go over our last method of charging, I want to discharge the electroscope and return it to its neutral state. I can accomplish this through **grounding**. Grounding is the process of providing a conductive pathway between a charged object and the ground, allowing the charged object to become neutral. In our example, the electroscope had an excess of electrons. If I touch the electroscope with my finger, the excess electrons can travel onto my finger, through my body, and into the ground. Grounding my electroscope will make it neutral again.

Induction is the process of charging a neutral object without ever having it touch another charged object. Induction involves bringing a charged object close enough to a neutral object to polarize it, and then grounding the neutral object in order to allow a charge to build up on it. Finally, the ground is removed before the original charged object is taken away.

So, unlike conduction, induction does not involve direct contact between the original charged object and the original neutral one. Another important difference between conduction and induction is that while conduction results in the neutral object taking on the same type of charge as the charged object, induction results in the neutral object taking on the opposite type of charge as the charged object.

Let's review before moving on.

Lesson 5–3 Review

1. ______________ is the process of charging an object by bringing it into direct contact with another charged object.
2. ______________ is the process of providing a conductive path between a charged object or a circuit to the ground.
3. ______________ is the process of inducing a temporary separation of charges in a neutral object by bringing it in close proximity to a charged object.

Lesson 5–4: Electric Fields

How can an object exert a force on another object when they aren't even touching? What is the nature of a force at a distance? The idea of an object interacting on a distant object is easier to visualize when we employ the model of force fields. In this model, you imagine each object surrounded by a field, or area. In the case of gravitational fields, the more massive the object is, the stronger the gravitational field it is surrounded by. In the case of electric fields, objects with greater charges (q) are surrounded by stronger electric fields.

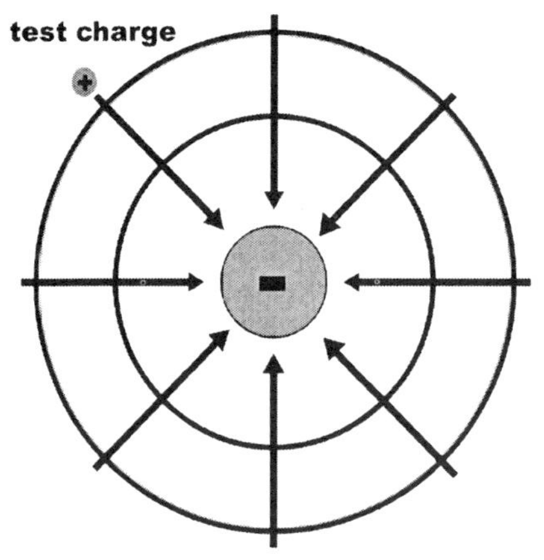

Figure 5.6

Not all electric fields are uniform in terms of strength. Just as the force of attraction to Earth due to gravity is greatest near its surface, the electrostatic force of attraction or repulsion for an ion is greater as you get closer to it. The **electric field strength** (E) at a given point in the field is measured in terms of how much force a charged object within it experiences at that position, per units of its charge. Electric field strength is also called **electric field intensity**.

Electric Field Intensity

$$\frac{\text{force experienced by a charged object at a point in the field}}{\text{the charge on this object}}$$

$$\text{or}\quad E = \frac{F}{q}.$$

The units used for measuring electric field intensity are newtons/coulomb (N/C).

Example 1

Calculate the magnitude of the force that a proton will experience if it is placed at a point in an electric field where the intensity is 0.15 N/C. (Recall that the charge on one proton is 1.60×10^{-19} C.)

Given: $E = 0.15$ N/C $\qquad q = 1.60 \times 10^{-19}$ C

Find: F

Isolate: $E = \frac{F}{q}$

Multiply both sides by q: $q \times E = \frac{F}{\cancel{q}} \times \cancel{q}$

Solution: $F = qE = (1.60 \times 10^{-19}\ \cancel{C})(0.15\ N/\cancel{C}) = \mathbf{2.4 \times 10^{-20}\ N}$

We treat tiny charged objects as *point charges*, meaning we treat them as if they take up no space and a uniform radial field surrounds them. Sometimes you won't be given the charge of a test charge or the force experienced by the test charge. In these situations you can combine the formula for electric field intensity with Coulomb's law to find the strength of the electric field generated by a point charge at a given distance.

Field intensity formula: $E = \frac{F}{q}$

Coulomb's law: $F = k_c \frac{q_1 q_2}{d^2}$

Substitute: $E = \dfrac{k_c \dfrac{q_1 q_2}{d^2}}{q}$

Simplify: $E = \dfrac{k_c \dfrac{q_1 \not{q_2}}{d^2}}{\not{q}} = \dfrac{k_c q}{d^2}$

Example 2

Determine the strength of the electric field at a distance of 3.0 × 10^{-3} m from an oxide (O^{2-}) ion.

Convert: Remember, an oxide (O^{2-}) ion has two extra electrons, so we find its net charge by multiplying the elementary charge (e) by 2.

$q = n \times e = 2\,\not{e^-} \times -1.60 \times 10^{-19}\ C/\not{e^-} = \mathbf{-3.20 \times 10^{-19}\ C}$

Because we are finding only the magnitude (strength) of the electric field, and not the direction, we can use the absolute value of the charge.

Given: $q = 3.20 \times 10^{-19}$ C $\quad d = 3.0 \times 10^{-3}$ m

$k_c = 9.0 \times 10^{9}\ N \cdot m^2/C^2$

Find: E

Solution: $E = \dfrac{k_c q}{d^2} = \dfrac{(9.0 \times 10^{9}\ N \cdot \not{m}^2/C^2)(3.20 \times 10^{-19})}{(3.0 \times 10^{-3}\ \not{m})^2} = \mathbf{3.2 \times 10^{-4}\ N/C}$

By measuring the strength of an electric field at many different spots, you can "map" it out in the same way that you can make a topographic map from many measurements of altitude. An electric field diagram is a qualitative representation of such a map, following these conventions:

1. Field lines are drawn as arrows, pointing in the direction of the force a positive test charge would experience at that point in the field. (For curved lines, the force experienced at that point is tangent to the curve.)

Fields around charged objects

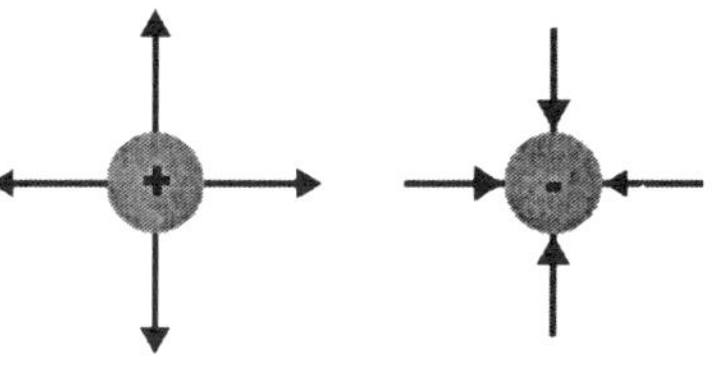

Figure 5.7

2. Field lines are never drawn crossing each other.
3. The more lines per given unit of area, the stronger the field at that point.

Not all charged objects are treated as point charges. A capacitor, which is a device that can be used to store charge, is made up of two metallic plates. When these plates are charged, one will have a net positive charge and the other will have a net negative charge. The type of field between the plates is called a uniform electric field.

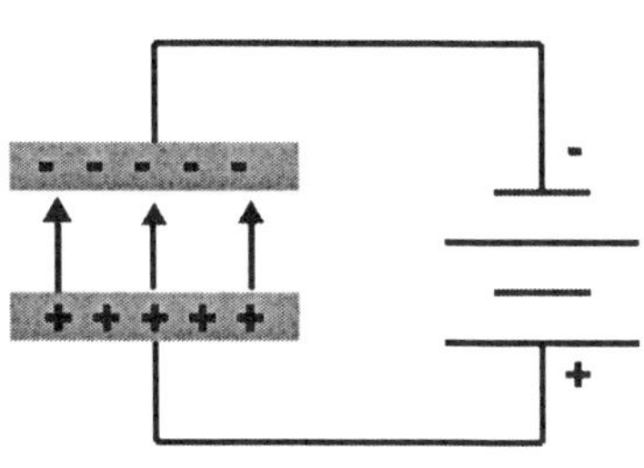

Figure 5.8

In the type of electric field formed between two charged plates, the electric potential (V) decreases linearly with the distance, not with the square of the distance, as in the field surrounding a point charge. The formula for calculating the average strength of a uniform electric field is:

$$E_{avg} = \frac{V}{d}$$

Where d is the distance between the plates, and V is the voltage, or potential difference between the plates.

Lesson 5–4 Review

1. How would the force on an electron in an electric field compare to the force experienced by a proton in the same location in the same electric field.
2. If an electron experiences an electrostatic force of -1.1×10^{-18} N in a field, calculate the electric field strength.
3. Calculate the magnitude of the electric field strength at a distance of 1.5×10^{-4} m from a proton.

Lesson 5–5: Electric Potential Energy

In Chapter 3 we discussed gravitational potential energy. We said that when a person lifts an object up, he or she is doing work against the gravitational field (or "gravity" for short) around Earth. At its new height, the object would have more gravitational potential energy, with reference to the floor. When an object falls, the gravitational field does work on the object. This work exerts a force over a distance, and the object's gravitational potential energy is changed into kinetic energy as it falls. If you can transfer this entire mental model over to the electric charges and force, you will be on your way to understanding electric potential energy.

It is really quite simple. **Electric potential energy** is the energy stored by an electric charge due to its position in an electric field. Imagine holding a positive test charge near a stationary negatively charged object. The electric field is directed towards the negatively charged object, indicating the direction of force that a positively charged object would experience within it. The positive test charge that we hold has a certain amount of electric potential energy at this point. If we release it, the field around the negative charge does work on our positive test charge, and attracts it. The electric potential energy of our test charge is converted into kinetic energy as it accelerates towards the negative charge. The test charge "falls" towards the negative charge, similar to the way that an apple falls toward Earth.

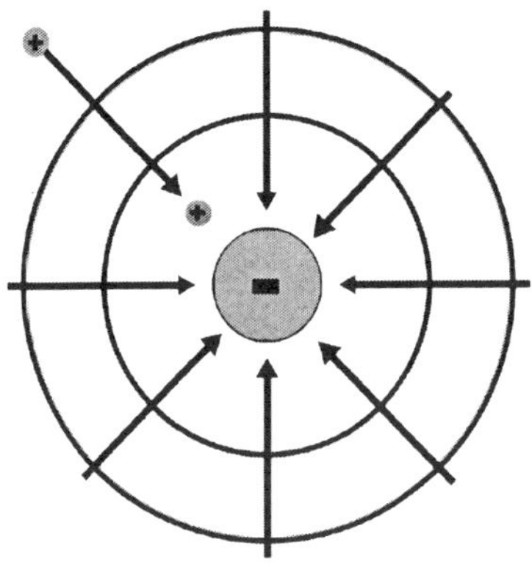

Figure 5.9

Of course, there is also a repulsive force associated with electric charges. If we hold our positive test charge near another positively charged object that is stationary, there will be a repulsive force between them. The test charge still has a certain amount of electric potential energy due to its position, but when we release the test charge, it will "fall" away from the other charged object. Its electric potential energy is converted to kinetic energy as it moves away from the other charged object.

The electric potential energy that a charged object possesses due to its proximity to another charged object is proportional to the product of the charges, and inversely proportional to the distance between them.

$$P.E._{electric} = k_c \frac{q_1 q_2}{d}$$

Example 1

Calculate the electric potential energy between two oxide (O^{2-}) ions that are separated by a distance of 4.8 × 10^{-3} m.

First, let's find the charge on each of the oxide ions, in coulombs:

$$q_1 = q_2 = n \times e = 2\,\cancel{e^-} \times -1.60 \times 10^{-19}\ \text{C}/\cancel{e^-} = \mathbf{-3.20 \times 10^{-19}\ C}$$

Given: $k_c = 9.0 \times 10^9\ \text{N} \cdot \text{m}^2/\text{C}^2$ $\quad q_1 = -3.20 \times 10^{-19}\ \text{C}$
$q_2 = -3.20 \times 10^{-19}\ \text{C}$ $\quad d = 4.8 \times 10^{-3}\ \text{m}$

Find: $P.E._{electric}$

Solution:

$$P.E._{electric} = k_c \frac{q_1 q_2}{d}$$

$$= \frac{(9.0 \times 10^9\ \text{N} \cdot \text{m}^2/\text{C}^2)(-3.20 \times 10^{-19}\ \text{C})(-3.20 \times 10^{-19}\ \text{C})}{(4.8 \times 10^{-3}\ \text{m})}$$

$$= \mathbf{1.9 \times 10^{-25}\ J}$$

In terms of the electronics we will be studying in future chapters, we are more concerned with uniform electric fields, such as the fields between two charged plates, than we are with the electric fields surrounding point charges. Unlike the fields around point charges, which decrease over distance, uniform electric fields show the same intensity at each location. The change in the electric potential energy associated with a charged object in a uniform electric field can be calculated with the following formula:

$$\Delta P.E._{electric} = -qEd$$

where q is the charge on the object, E is the strength of the electric field, and d is the displacement of the charged object in the direction of the field. Electric potential energy is measured in joules (J).

It is important to remember that the rules of work apply here. There is only a change in electrical potential energy if work is done on or by the electric field. In the same way that there is no change in the gravitational potential energy of a bowling ball if you simply slide it to the right side of a shelf, without changing its height, there is no change in the electrical potential energy of a charged particle if it doesn't change its position relative to the reference point in the electric field. When the displacement is in the same direction of the electric field, we can use the formula for finding the change in electric potential energy, because the cosine of a zero degree angle is equal to 1. However, when the direction that the charged particle moves is anything other than parallel to the field, we really need to include the cosine of the angle between the field and the motion in our formula, as shown here:

$$\Delta P.E._{electric} = -qEd\cos\theta .$$

Example 2

Calculate the change in electrical potential energy of a proton when it moves 3.7 cm parallel to the direction of a uniform electric field with a strength of 2.5 N/C.

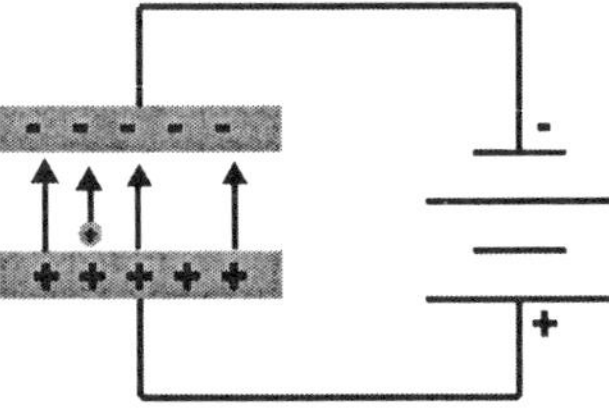

Figure 5.10

Convert: $3.7\ \text{cm} \times \frac{1\ \text{m}}{100\ \text{cm}} = \mathbf{0.037\ m}$

Note that the distance the proton travels is parallel to the electric field, and cos 0 = 1, so we don't need to include the angle in our calculation.

Given: $q = 1.60 \times 10^{-19}$ C E = 2.5 N/C d = 0.037 m

Find: $\Delta P.E._{electric}$

Solution:
$$\Delta P.E._{electric} = -qEd = -(1.60\times 10^{-19}\ \text{C})(2.5\ \text{N/C})(0.037\ \text{m})$$
$$= \mathbf{-1.5\times 10^{-20}\ J}$$

You can see how the negative sign in the formula comes into play. As the proton moves in the direction of the electric field, it is analogous to an object falling towards Earth. Each of these motions results in a loss of potential energy.

Example 3

An ion of oxide (O^{2-}) moves through a uniform electric field with a strength of 1.4 N/C. It travels 45 cm in a straight line at an angle of 127.0° from the direction of the field. Calculate the change in the electric potential energy of the ion.

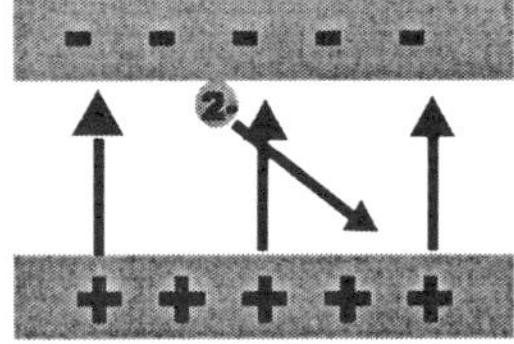

Figure 5.11

Convert: $45\ \not{cm} \times \frac{1\text{ m}}{100\ \not{cm}} = \mathbf{0.45\ m}$

The oxide ion has two extra electrons, so

$q = n \times e = 2\ \not{e}^- \times -1.60 \times 10^{-19}\ \text{C}/\not{e}^- = \mathbf{-3.20 \times 10^{-19}\ C}$

Given: $q = -3.20 \times 10^{-19}$ C　　$E = 1.4$ N/C
$d = 0.45$ m　　$\theta = 127.0°$

Find: $\Delta P.E._{electric}$

Solution:

$$\Delta P.E._{electric} = -qEd\cos\theta$$
$$= -(-3.20 \times 10^{-19}\ \not{C})(1.4\ \text{N}/\not{C})(0.45\ \text{m})(\cos 127.0°)$$
$$= \mathbf{1.2 \times 10^{-19}\ J}$$

Notice that our ion in Example 2 shows an increase in electric potential energy, indicated by the implied positive sign of our answer. Where did this extra energy come from? Just as a man does work on a bowling ball to raise it to a greater height, work must have been done on the oxide ion to move it to a position where it possesses more electrical potential energy. The formula below summarizes the relationship between work and the change in electric potential energy.

$$\Delta P.E._{electric} = -W$$

The negative sign for work in our formula may be easier to understand if you look at our two examples. In the first example, work was done *by* the field (+W), resulting in a negative change in electric potential energy. In our second example, work was done on the field (–W), and the change in electric potential energy was positive.

Combining the two formulas from this lesson, you can see that

$\Delta P.E._{electric} = -W = -qEd\cos\theta$.

Consequently, $W = qEd\cos\theta$.

Example 4

How much work would be required to move a proton ($q = 1.60 \times 10^{-19}$ C) a distance 0.85 m directly against a uniform electric field with a strength of 5.6 N/C?

Note: The phrase "directly against a…field" is meant to imply that the angle between the displacement and the field is 180°.

Given: $q = 1.60 \times 10^{-19}$ $E = 5.6$ N/C $d = 0.85$ m $\theta = 180°$

Find: W

Solution: The work done *by* the field is:

$$W = qEd\cos\theta = (1.60 \times 10^{-19})(5.6 \text{ N/C})(0.85 \text{ m})(\cos 180°)$$
$$= -7.6 \times 10^{-19} \text{ J}$$

So, the magnitude of the required work done *on* the field is **7.6 × 10⁻¹⁹ J.**

Electric Potential

One of the factors that may contribute to the trouble that some students have with learning about electricity and magnetism is the fact that some of the terms that are used sound so similar. We just finished discussing *electric potential energy*, and in the next lesson we will discuss *potential difference*, but first, we will discuss **electric potential**. Electric potential sounds like electric potential energy, and although these concepts are related, they are not exactly the same thing. Electric potential is the electric potential energy that a charged object has, divided by its charge. The formula for calculating the electric potential of a point charge is:

$$\text{electric potential} = \frac{\text{electric potential energy}}{\text{charge}}.$$

Lesson 5–5 Review

1. _______________ is the energy that a charged object has due to its location in an electric field.

2. Calculate the electric potential energy between two protons separated by a distance of 2.8×10^{-4} m.

Lesson 5–6: Potential Difference

A bowling ball on a shelf has more gravitational potential energy sitting on a high shelf than sitting on a low shelf. In much the same way, a charged particle will have more electric potential energy at one point in an electric field than it will have in another. If we take the difference in the electric potential energy that a test charge has between two points in the electric field and then divide by the charge on our test charge, we get the quantity known as **potential difference**, which is often called **voltage**.

$$\text{potential difference}\ (\Delta V) = \frac{\text{change in electric potential energy}\ (\Delta P.E._{\text{electric}})}{\text{charge}\ (q)}$$

or

$$\Delta V = \frac{\Delta P.E._{\text{electric}}}{q}$$

Because electric potential energy is measured in joules and charge is measured in coulombs, potential difference is measured in J/C. An equivalent derived unit called the volt (V) has been introduced.

$$1\ \text{Volt}\ (V) = \frac{1\ \text{Joule}\ (J)}{1\ \text{Coulomb}\ (C)}$$

Example 1

Find the change in electrical potential energy as a proton moves through a potential difference of 1.5 V.

Given: $q = 1.60 \times 10^{-19}$ C $\qquad \Delta V = 1.5$ V

Find: $\Delta P.E._{\text{electric}}$

Isolate:

Starting with the original formula: $\Delta V = \dfrac{\Delta P.E._{\text{electric}}}{q}$

We multiply both sides by q: $q \times \Delta V = \dfrac{\Delta P.E._{\text{electric}}}{\not{q}} \times \not{q}$

Rearranging the equation, we get: $\Delta P.E._{\text{electric}} = q\Delta V$

Solution: $\Delta P.E._{\text{electric}} = q\Delta V = (1.60 \times 10^{-19}\ C)(1.5\ V) = \mathbf{2.4 \times 10^{-19}\ J}$

Our answer for Example 1 is so small that I'm sure you'll understand why the unit called the **electron volt** (eV) was introduced. An electron volt is defined as the energy required to move an electron between two points that have a difference of potential of 1 volt. We can find the value of 1 electron volt using the same equation that we used for Example 1.

$$\Delta P.E._{electric} = q\Delta V = (1.60 \times 10^{-19}\ C)(1.0\ V) = \mathbf{1.6 \times 10^{-19}\ J}$$

So, 1 eV = 1.6×10^{-19} J

If we were asked to convert our answer for Example 1 into electron volts, we could convert as shown here:

$$2.4 \times 10^{-19}\ \mathbf{J} \times \frac{1\ eV}{1.60 \times 10^{-19}\ \mathbf{J}} = 1.5\ eV\ .$$

Of course, this answer makes perfect sense when you think about it. The proton has the same charge as an electron, and the potential difference in Example 1 was 1.5V, so we would need 1.5 eV to move the proton.

When we talk about batteries or cells in terms of *voltage*, we are really talking about the difference in potential between the positive and negative terminals. When we discuss circuits in the next chapter, you will see that we often talk about the voltage or potential difference "across" a component. In these situations we are talking about the difference in potential between a point before the current enters the component and at another point after the current leaves the component.

Combining the equation for potential difference with the relationship between electric potential energy from last lesson, $\Delta P.E._{electric} = -W$, we can see how work relates to potential difference.

$$\text{potential difference } (\Delta V) = \frac{\text{work (W)}}{\text{charge (q)}}$$

or

$$\Delta V = \frac{W}{q}$$

Example 2

How much work is required to move a charge of 3.0 coulombs across a potential difference of 9.0 V?

Given: q = 3.0 C ΔV = 9.0 V

Find: W

Isolate:

Starting with the equation: $\Delta V = \frac{W}{q}$

We multiply both sides by q: $q \times \Delta V = \frac{W}{\cancel{q}} \times \cancel{q}$

Rearranging the equation, we get: $W = q\Delta V$

Solution: $W = q\Delta V = (3.0\,C)(9.0\,V) = \mathbf{27\,J}$

Lesson 5–6 Review

1. ______________ the energy required to move an electron between two points that have a difference of potential of 1 volt.
2. Find the change in electrical potential energy, in electron volts, as a proton moves through a potential difference of 4.5.
3. An alpha particle has a charge of 3.20×10^{-19} C. If it takes 3.84×10^{-18} J to move the alpha particle between two points in an electric field, find the potential difference between the two points.

Lesson 5–7: Capacitance

When I was a young teacher, with perhaps three years of experience, a couple of other teachers and I brought a group of students on a trip to Washington, D.C. At one point the students had some time to go shopping at the mall. Later, at dinner, one of the students wanted to show me the "novelty" lighter he had purchased. Some of the other students warned me not to press the button on the lighter, explaining that it produced a shock. I had experience with another novelty toy, the "joy buzzer," which also supposedly produced a shock, but really only produced a mild vibration instead. Expecting a similar painless "shock," I confidently pressed the button on the novelty lighter. Well, my whole arm instantly went numb, and my hand dropped as I lost control of the muscles in my arm! The lighter actually contained a powerful **capacitor**, capable of delivering a very real and painful shock!

Capacitors are devices that can be used to store energy in the form of separated charges. A typical capacitor consists of two parallel metallic plates, separated by an insulating material called a dielectric. When the capacitor is connected to a source of potential difference, one plate takes on a net positive charge and the other takes on a net negative charge, and the capacitor is said to be "charged." The charged state of a capacitor will only last until a connection is made, which allows the charges to flow again. When a connection is made between the two plates of a capacitor, an electric current results, as the charges flow to re-establish equilibrium and the capacitor is discharged. As I was reminded from my run-in with the novelty lighter, a charged capacitor can be a source of potential energy.

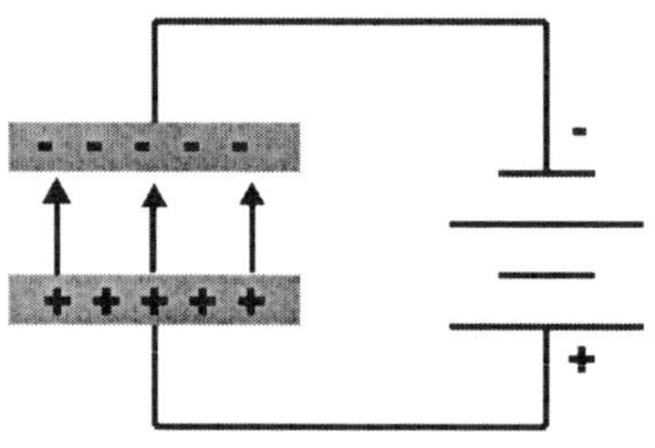

Figure 5.12

Potential energy stored in a capacitor = $\frac{1}{2}$ charge on one plate × the potential difference

$$P.E._C = \frac{1}{2}Q\Delta V$$

Example 1

Calculate the electrical potential energy stored in a capacitor with a magnitude of charge of 4.5 μC on each plate, and a potential difference of 1.5 V across its plates.

Convert: The symbol μ stands for "micro." 1C = 1 × 10^6 μC.

$$4.5\,\mu C \times \frac{1\,C}{1\times 10^6\,\mu C} = 4.5\times 10^{-6}\,C$$

Given: $Q = 4.5 \times 10^{-6}$ C $\Delta V = 1.5$ V

Find: $P.E._c$

Solution: $P.E._c = \frac{1}{2}Q\Delta V = (0.5)(4.5 \times 10^{-6}\,C)(1.5\,V) = \mathbf{3.4 \times 10^{-6}\,J}$

The flash in a standard camera is operated by a capacitor. The battery in the camera charges the capacitor. When a picture is taken, the capacitor discharges through the flashbulb, allowing it to get quite bright for a short time. The capacitor needs a brief period to recharge as the battery separates the charges on the plates once more.

A defibrillator, which you have probably seen on TV, if not in real life, also makes use of capacitors. The medical professional holds the two paddles of the defibrillator together until they are properly charged. He or she then yells, "Clear!" and applies the paddles to the body of the patient. A quick discharge sends electrical current into the body of the patient in an attempt to cause the heart to resume beating normally.

The ability of a capacitor to store a large amount of charge per volt is called **capacitance**.

$$\text{Capacitance (C)} = \frac{\text{The charge on one of the capacitor's plates (Q)}}{\text{The potential difference across the two plates } (\Delta V)}$$

or

$$C = \frac{Q}{\Delta V}$$

Capacitance is measured in units called farads (f), where 1 farad is equal to 1 coulomb/volt.

Example 2

Each plate on a 6.0 μf capacitor has a charge with a magnitude of 1.2×10^{-3} C on it. What is the potential difference across its plates?

Convert: The symbol μ stands for "micro." $1\text{f} = 1 \times 10^{6}\mu\text{f}$

$$6.0\ \mu\text{f} \times \frac{1\ \text{f}}{1 \times 10^{6}\ \mu\text{f}} = \mathbf{6.0 \times 10^{-6}\ f}$$

Given: $Q = 1.2 \times 10^{-3}$ C $\quad C = 6.0 \times 10^{-6}$ f

Find: ΔV

Solution: $$\Delta V = \frac{Q}{C} = \frac{1.2 \times 10^{-3}\ \text{C}}{6.0 \times 10^{-6}\ \text{f}} = \mathbf{2.0 \times 10^{2}\ V}$$

If we take our last formula and isolate charge (Q), we get:

$Q = CV.$

We can take this value for Q and substitute it into the formula we used to solve Example 1 to derive another useful formula.

$$P.E._C = \frac{1}{2}QV = \frac{1}{2}CV \times V = \frac{1}{2}CV^2$$

Example 3

How much energy is stored in a 3.5 μf capacitor when it is connected to a 3.0 V battery?

Convert: $3.5\,\mu f \times \frac{1\,f}{1\times 10^6\,\mu f} = \mathbf{3.5 \times 10^{-6}\ f}$

Given: $C = 3.5 \times 10^{-6}$ f $V = 3.0$ V

Find: $P.E._c$

Solution: $P.E._C = \frac{1}{2}CV^2 = (0.5)(3.5\times 10^{-6}\ f)(3.0\ V)^2 = \mathbf{1.6 \times 10^{-5}\ J}$

The capacitance of a capacitor is based on several factors, including the material of the plates, the dielectric material in between the plates, the size and shape of the plates, and the distance between the plates. You should remember that the capacitance of a capacitor is directly proportional to the area of one of the metal plates, and inversely proportional to the distance between them.

$$\text{capacitance} \propto \frac{\text{area of one of the metal plates}}{\text{distance between the plates}}$$

or

$$C \propto \frac{A}{d}$$

Example 4

If you double the area of both plates of a capacitor, and then double the distance between the plates, what will happen to the capacitance of the capacitor?

Solution: $C \propto \frac{2A}{2d}$

You would be multiplying both the denominator and numerator by 2, so the capacitance will remain the same.

A proportionality constant called the *permittivity of free space*, or vacuum permittivity (ε_0), can be used to set the sides of the equation equal to each other.

$$C = \frac{\varepsilon_0 A}{d}$$

$\varepsilon_0 = 8.85 \times 10^{-12}\ C^2/N \cdot m^2$

Example 5

A capacitor is made up of two parallel plates separated by a distance of 0.004 m with a vacuum in between them. The area of each plate is 0.050 m². Find the capacitance of this capacitor.

Given: $\varepsilon_0 = 8.85 \times 10^{-12}\ C^2/N \cdot m^2$ $A = 0.050\ m^2$ $d = 0.004\ m$

Find: C

$$C = \frac{\varepsilon_0 A}{d} = \frac{(8.85 \times 10^{-12}\ C^2/N \cdot m^2)(0.050\ m^2)}{0.004\ m} = \mathbf{1 \times 10^{-10}\ f}$$

The permittivity of air is so close to that of vacuum that we use the same value for both. When a different substance, such as paper, is used as the dielectric between the plates of the capacitor, you will need to multiply the permittivity of free space by the dielectric constant (k) for that particular material, making the formula:

$$C = \frac{k\varepsilon_0 A}{d}$$

Lesson 5–7 Review

1. If you increase the distance between the charged plates of a capacitor, what will happen to its capacitance?
2. Find the capacitance of a capacitor with a potential difference of 9.0 V across its plates when each plate has a charge with a magnitude of 1.6 μC.

3. Calculate the electrical potential energy stored in a capacitor with a magnitude of charge of 2.5 μC on each plate, and a potential difference of 6.0 V across its plates.

Chapter 5 Examination

Part I—Matching

Match the following terms to the definitions that follow.

a. electroscope d. positive ion g. grounding

b. induction e. negative ion h. capacitor

c. conduction f. polarization i. electric potential

_____1. A device that stores an electric charge.

_____2. The process of charging an object by bringing it near a charged object and providing a pathway for charge between the neutral object and the ground.

_____3. An atom that has gained one or more electrons.

_____4. The electric potential energy of a charged object, divided by its charge.

_____5. Providing a conductive path between a charged object or a circuit to the ground.

Part II—Multiple Choice

For each of the following questions choose the best answer.

6. A neutral atom must have an equal number of
   a) protons and electrons
   b) protons and neutrons
   c) neutrons and electrons
   d) protons, neutrons, and electrons

7. An electroscope obtains a charge of -1.44×10^{-16} C. How many excess electrons does this represent? ($e = -1.60 \times 10^{-19}$ C)
   a) $2.30 \times 10^{-35}\ e^-$
   b) $1.44 \times 10^{-16}\ e^-$
   c) $1.2 \times 10^{3}\ e^-$
   d) $9.00 \times 10^{2}\ e^-$

8. A metal sphere with an excess of 17 electrons touches an identical metal sphere with an excess of 5 electrons. After the spheres are separated, how many excess electrons would each sphere have?
   a) 5 b) 11 c) 12 d) 22

9. The electrostatic force of attraction between two charged objects is F. What would be the force between the objects if the charge on each of the objects were tripled?
   a) 3F b) –3F c) 6F d) 9F

10. The electrostatic force of attraction between two charged objects separated by a distance of 1.0 cm is given by F. If the distance between the objects were increased to 5.0 cm, what would be the electrostatic force of attraction between them?
   a) 5F b) $\frac{F}{5}$ c) 25F d) $\frac{F}{25}$

11. What is the magnitude of the force experienced by an electron in an electric field with an intensity of 3.50×10^5 N/C?
   a) 2.19×10^{24} N
   b) 5.60×10^{-14} N
   c) 1.9×10^{22} N
   d) 4.5×10^{-14} N

12. Calculate the electric field intensity at a distance of 0.875 m away from a point charge of 1.70 C. ($k_c = 9.0 \times 10^9\ \text{N} \cdot \text{m}^2/\text{C}^2$)
   a) 1.9 N/C
   b) 1.7×10^{10} N/C
   c) 2.0×10^{10} N/C
   d) 1.3×10^9 N/C

13. 5.8×10^{-19} J of work are done by a uniform electric field with a strength of 3.9 N/C in order to move an electron. Find the distance the electron travels.
   a) 0.82 m
   b) 1.3 m
   c) 1.3×10^{-19} m
   d) 23 m

14. 12.0 J of work are required to move a 4.0 C charge between two points in an electric field. Calculate the potential difference between these two points.
   a) 3.0 V b) 8.0 V c) 16.0 V d) 48 V

Part III—Calculations

Perform each of the following calculations.

15. A 3.50 μf capacitor has a charge with a magnitude of 4.50 μC on each parallel plate. Find the potential difference between the plates.
16. A capacitor stores 12.0 J of potential energy when the potential difference between the parallel plates is 9.0 V. Find the capacitance of the capacitor.
17. Two charged objects with charges of 2.5 μC and 0.45 μC, respectively, experience an electrostatic force of repulsion of 1.4×10^{-4} N. What is the distance between the objects?
18. Find the intensity of an electric field in which an electron experiences a force of 6.0 N.
19. How much work would be required to move a charge of 4.0 C between two points in a circuit with a potential difference of 9.0 V?
20. An atom of oxygen contains 8 protons. What is the nuclear charge of an oxygen atom in coulombs?

Answer Key

The actual answers will be shown in brackets, followed by an explanation. If you don't understand an explanation that is given in this section, you may want to go back and review the lesson that the question came from.

Lesson 5–1 Review

1. [a proton]—the proton has a charge that is equal in magnitude to the charge on an electron, but with an opposite sign.
2. [a negative ion]—A negative ion has gained additional negative charges, giving it a net charge that is negative.
3. [-3.20×10^{-19} C]—We simply multiply 2 by the charge on 1 electron.

   $q = n \times e = 2\,¢^{-}\,(-1.60 \times 10^{-19}\ \text{C}/¢^{-}) = -3.20 \times 10^{-19}\ \text{C}$

Lesson 5–2 Review

1. [Coulomb's law]—The formula for Coulomb's law is: $F = k_c \dfrac{q_1 q_2}{d^2}$.

2. [The force would be quadrupled.]—Doubling the charge on each object would cause the product of the charges to be multiplied by a factor of (2 × 2) 4.
3. [The force would be 1/4th as great.]—Because the force is inversely proportional to the square of the distance between the two objects, doubling the distance will cause the force to be divided by a factor of (2^2) 4.

Lesson 5–3 Review

1. [conduction]—It may help to remember the letter c in both *c*onduction and *c*ontact.
2. [grounding]—Grounding a charged object will eliminate the excess charge and make it neutral.
3. [polarization]—When you bring a charged balloon close to a neutral wall, the wall becomes polarized, as the electrons near the surface of the wall are repelled by the excess electrons on the balloon.

Lesson 5–4 Review

1. [The forces would be equal in magnitude, but opposite in direction.]—The formula F = qE shows that the force experienced by a particle will be directly proportional to its charge. The proton and electron have equal but opposite charges, so the forces that they experience will be equal and opposite as well.
2. [6.9 N/C]—As long as you know the charge on an electron, -1.60×10^{-19} C, the rest is easy.

$$E = \frac{F}{q} = \frac{(-1.1 \times 10^{-18}\ \text{N})}{(-1.60 \times 10^{-19}\ \text{C})} = 6.9\ \text{N/C}$$

3. [0.064 N/C]— $E = \dfrac{k_c q}{d^2} = \dfrac{(9.0 \times 10^9\ \text{N} \cdot \cancel{\text{m}}^2/\text{C}^2)(1.60 \times 10^{-19})}{(1.5 \times 10^{-4}\ \cancel{\text{m}})^2} = 0.064\ \text{N/C}$

Lesson 5–5 Review

1. [electric potential energy]—For point charges, we can find the electric potential energy with the formula:

$$\text{P.E.}_{\text{electric}} = k_c \frac{q_1 q_2}{d}.$$

For charged plates, we use: $\Delta\text{P.E.}_{\text{electric}} = -qEd$.

2. [8.2×10^{-25} J]—

$$\text{P.E.}_{\text{electric}} = k_c \frac{q_1 q_2}{d} = \frac{(9.0 \times 10^9\ \text{N} \cdot \text{m}^2/\text{C}^2)(1.60 \times 10^{-19}\ \text{C})(1.60 \times 10^{-19}\ \text{C})}{(2.8 \times 10^{-4}\ \text{m})}$$

$$= 8.2 \times 10^{-25}\ \text{J}$$

Lesson 5–6 Review

1. [electron volt]—Although this term includes the word "volt", it is really a unit of energy. The conversion between electron volts and joules is:
   1 eV = 1.6 × 10^{-19} J.
2. [4.5 eV]—One proton has the same charge as one electron, so we will use 1.00 e as the value for q. $\Delta P.E._{electric} = q\Delta V = (1.00\text{ e})(4.5\text{ V}) = 4.5\text{ eV}$
3. [12.0 V]—$\Delta V = \frac{W}{q} = \frac{3.84 \times 10^{-18}\text{ J}}{3.20 \times 10^{-19}\text{ C}} = 12.0\text{ V}$

Lesson 5–7 Review

1. [The capacitance will decrease]—The capacitance of a capacitor is inversely proportional to the distance between its plates, as shown by the formula:
   $C = \frac{\varepsilon_0 A}{d}$
2. [1.8 × 10^{-7} f]—Remember to change 1.6 μC into 1.6 × 10^{-6} C, and then solve:
   $C = \frac{Q}{\Delta V} = \frac{1.6 \times 10^{-6}\text{ C}}{9.0\text{ V}} = 1.8 \times 10^{-7}\text{ f}$.
3. [7.5 × 10^{-6} J]—Remember to convert 2.5 μC to 2.5 × 10^{-6} C. Then solve:
   $P.E._c = \frac{1}{2} Q\Delta V$ = (0.5)(2.5 × 10^{-6} C)(6.0 V) = 7.5 × 10^{-6} J.

Chapter 5 Examination

1. [h. capacitor]—An electroscope is a device that can become charged, but it doesn't store the charge.
2. [b. induction]—The process of induction produces a charged object with a charge that is the opposite to the original charged object.
3. [e. negative ion]—Remember, it has gained additional negative charges (electrons), giving it a net negative charge.
4. [i. electric potential]—Voltage is the difference in electric potential between two points.
5. [g. grounding]—Lightning rods are based on this principle.
6. [a. protons and electrons]—If an atom has the same number of positive and negative charges, the net charge is zero.
7. [d. 9.00 × 10^2 e^-]— number of $e^- = \frac{q}{e} = \frac{-1.44 \times 10^{-16}\text{ C}}{-1.60 \times 10^{-19}\text{ C}/e^-} = 9.00 \times 10^2\ e^-$

8. [b. 11]—Remember, the excess electrons repel each other, so they will move to maximize the distance between them. The spheres are metal, so the electrons are free to spread out. The spheres are identical, so they will be evenly distributed over the surface of both spheres, resulting in the same number of excess electrons on each sphere.

$$\text{number of electrons on each sphere} = \frac{\text{total number of excess electrons}}{\text{total number of spheres}}$$
$$= \frac{(17+5)}{2} = 11$$

9. [d. 9F]—Coulomb's law tells us that the electrostatic force between two charged objects is directly proportional to the product of the charges, as shown by $F = k_c \frac{q_1 q_2}{d^2}$.

   If the charge on each object was tripled, the product of the charges would increase by a factor of (3 × 3) 9, giving us 9 × F.

10. [d. $\frac{F}{25}$]—Coulomb's law tells us that the electrostatic force between two charged objects is inversely proportional to the *square* of the distance between them, as shown by the equation

   $F = k_c \frac{q_1 q_2}{d^2}$.

   If the distance between the objects was increased from 1.0 cm to 5.0 cm, it was increased by a factor of 5. That means that the electrostatic force between the objects would decrease by a factor of $5^2 = 25$.

11. [b. 5.60×10^{-14} N]—You will need to recall that the charge on an electron is -1.60×10^{-19} C. Notice that we can figure out what operation to carry out by looking at the units, even if we couldn't recall the proper formula. The question asks for a force, which is measured in newtons, and we have one given (field intensity) measured in N/C and another given measured in C. The only way to get the coulombs to cross out is to multiply our givens together. $N/\cancel{C} \times \cancel{C} = N$. Of course, it is always better to look up the formula, if you have the opportunity.

   $F = Eq = (3.50 \times 10^5\ N/\cancel{C})(-1.60 \times 10^{-19}\ \cancel{C}) = -5.60 \times 10^{-14}\ N$

   The question asks only for the magnitude of the force, so that is why we leave the negative sign off our final answer.

12. [c. 2.0 × 10^10 N/C]—$E = \frac{k_c q}{d^2} = \frac{(9.0 \times 10^9\ N \cdot m^2/C^2)(1.70\ C)}{(0.875\ m)^2} = 2.0 \times 10^{10}\ N/C$

13. [a. 0.82 m]—$d = \frac{W}{qE\cos\theta} = \frac{5.10 \times 10^{-19}\ J}{(1.60 \times 10^{-19}\ C)(3.9\ N/C)(\cos 0^\circ)} = 0.82\ m$

14. [a. 3.0 V]—$\Delta V = \frac{\Delta P.E._{electric}}{q} = \frac{W}{q} = \frac{12.0\ J}{4.0\ C} = 3.0\ V$

15. [1.29 V]—$\Delta V = \frac{Q}{C} = \frac{4.50 \times 10^{-6}\ C}{3.50 \times 10^{-6}\ F} = 1.29\ V$

16. [0.30f]—$C = \frac{2P.E._c}{\Delta V^2} = \frac{2(12.0\ J)}{(9.0\ V)^2} = 0.30\ f$

17. [8.5 m]—

$$d = \sqrt{\frac{k_c q_1 q_2}{F}}$$

$$= \sqrt{\frac{(9.0 \times 10^9\ N \cdot m^2/C^2)(2.5 \times 10^{-6}\ C)(0.45 \times 10^{-6}\ C)}{1.4 \times 10^{-4}\ N}} = 8.5\ m$$

18. [3.8 × 10^19 N/C]—To solve this, recall that the charge on an electron is 1.60 × 10^-19 C. $E = \frac{F}{q} = \frac{6.0\ N}{1.60 \times 10^{-19}\ C} = 3.8 \times 10^{19}\ N/C$

19. [36 J]—$W = \Delta P.E. = q\Delta V = (4.0\ C)(9.0\ V) = 36\ J$

20. [1.28 × 10^-18 C]—$q = ne = 8(+1.60 \times 10^{-19}\ C) = +1.28 \times 10^{-18}\ C$

1 2 3 4 5 6 7 8 9 10

Electric Current and Circuits

When I was about 9 years old, my older (and much smarter) brother said, "Anyone who can't make a flashlight by the time that they are 9 years old must not be very smart." I remember feeling embarrassed because I didn't know how to make a flashlight, and I wondered if that meant that I was mentally deficient. I realize now that it wasn't that I was stupid, it was that my brother was exceptional (he went on to earn a Ph.D. in electrical engineering), and intelligence, like everything in physics, is relative. He did have a point, however. Flashlights are fairly easy to construct. They are slightly harder to understand. I hope that by the time you finish reading this chapter, you will be able to construct a flashlight, and understand how it, and other simple circuits work. Just don't use this knowledge to make *your* younger brother or sister feel inadequate.

Lesson 6–1: Current

The bulb of a flashlight will shine if there is a proper **electrical current** flowing through it. Electrical current is a measure of the rate of flow of charge through an object. It should be easy to remember if you think of how electrical current is analogous to water current.

$$\text{water current} = \text{rate of water flow} = \frac{\text{amount of water}}{\text{change in time}}$$

$$\text{electric current (I)} = \text{rate of charge flow} = \frac{\text{amount of charge (q)}}{\text{change in time }(\Delta t)}$$

If you measured the rate of water current, you would end up with units of volume over units of time. For example, a water current of 45 liters/minute would mean that 45 liters of water flow by a given point every minute. When we measure electrical current, we get units of charge over units of time. For example, an electrical current of 3.0 coulombs/second would mean that 3 coulombs of charge pass by a given area every second. The ampere is an SI base unit that is equivalent to an electrical current of 1 coulomb/second.

$$1\text{ ampere (A)} = \frac{1\text{ coulomb (q)}}{1\text{ second (s)}}$$

Example 1

How much charge, in coulombs, passes through a cross section of a particular wire every second if the current in the wire is 3.0 A? How many electrons (q of an e^- = 1.60×10^{-19} C) would this charge be equivalent to?

Given: I = 3.0 A Δt = 1.0 s

Find: q and number of e^-

Isolate:

We start with: $I = \frac{q}{\Delta t}$

Multiply both sides by Δt: $\Delta t \times I = \frac{q}{\cancel{\Delta t}} \times \cancel{\Delta t}$

We get: $q = I\Delta t$

Solve: $q = I\Delta t = (3.0\text{ A})(1.0\text{ s}) = (3.0\text{ C}/\cancel{\text{s}})(1.0\ \cancel{\text{s}}) = \mathbf{3.0\ C}$

To see how many electrons this charge would be equivalent to, we simply divide by the elementary charge (e), the charge on 1 electron.

$$\text{number of } e^- = \frac{q}{e} = \frac{3.0\ \cancel{\text{C}}}{1.60 \times 10^{-19}\ \cancel{\text{C}}/e^-} = \mathbf{1.9 \times 10^{19}\ electrons/second}$$

What did I mean in the first paragraph of this lesson when I said that a bulb required a "proper" current to light up? If the current is too small, that is, if not enough charge goes by per unit of time, the bulb will not light up. If the current is too great, meaning the rate of charge flow is greater than the bulb can handle, the bulb will break.

You may have heard of direct current (DC) and alternating current (AC), but not really understood what the terms mean. For example, in some situations, water flows in one direction only. If you turn on your sink, the water flows downwards. You can watch the running water for hours and never see the water flow upwards, or back into the pipe. This example is analogous to the type of electrical current called **direct current**. In direct current, the net charge travels in one general direction, although the individual charges don't move in straight lines. This could also be analogous to a river. The individual water molecules in a river move in random directions, but the net flow of water will always be downstream.

Alternating current would be more like the water you see at the beach. As you stand near the shore, the water from a wave moves up the beach to cover you feet, and then it recedes back. In alternating current, the net charge changes direction over and over as the net force on them is continuously reversed.

When you look at a river, it is easy to see the direction that the water flows. In what direction do charges in a wire, or charges in a solution, flow? It is important to remember that the current can represent any types of charges in motion. In solutions containing electrolytes, the dissolved ions would be the charge carriers. In wires, the electrons are the charge carriers. However, when people use the term "electric current," they are almost always referring to what is known as **conventional current**. Conventional current is always in the direction of the net motion of positive charge. In a simple flashlight, conventional current is in the opposite direction as the flow of electrons.

Now that we have gone over the concept of current, let's discuss the components that you will need to construct a flashlight. You will need to take a couple of simple items and set them up in what is called a **circuit**. An electrical circuit is an arrangement of electrical components that provide a continuous pathway for electricity to flow through. Some circuits, such as our flashlight, are very simple. Others are much more complex. The first thing that you need in order to construct your flashlight is a source of potential difference, such as a battery. The two terminals of the battery represent sites of varying electric potential energy. The electrons "fall" from the negative terminal, through the components of the circuit, back to

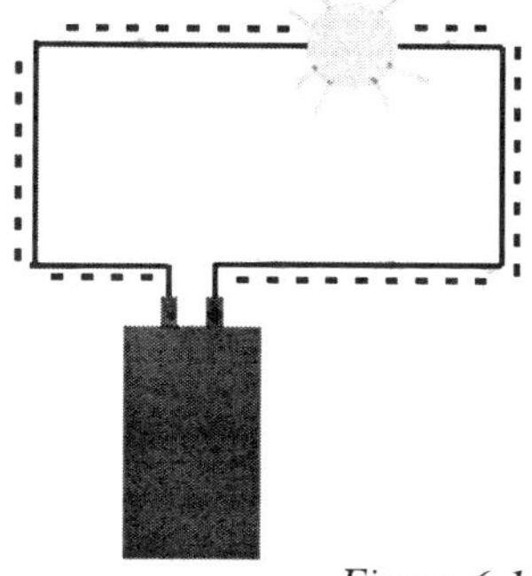

Figure 6.1

the positive terminal. The second thing that you need is a bulb. As the electrons travel through the filament of the bulb, some of the electric potential energy is transferred into heat and light. The final component is a conductor, such as wire, to attach the battery to the bulb.

A battery, a bulb, and some wire are all it takes to construct a simple flashlight. When you get past the terminology involved and the abstract concept of electron flow, a flashlight really is easy to make. In the remaining lessons in this chapter, I will explain more about the concepts involved in circuit design.

Lesson 6–1 Review

1. ______________ is electric current that flows in one direction.
2. A wire carries a current of 2.0 A. How much charge passes by a point in 55.0 seconds?
3. 175 C of charge pass a point in a wire in 35 s. What is the current in the wire?

Lesson 6–2: Resistance

When we discussed the construction of our imaginary flashlight in our last lesson, we said that the bulb converts some of the kinetic energy of the charges in motion into heat and light. The reason this happens is because the filament used inside the bulb offers more resistance to the motion of the charges than the copper wire that is used to connect the battery to the bulb. In physics, **resistance** is defined as the opposition to the flow of charges through a conductor. If you rub your hands vigorously together, you notice that they get hotter. In fact, you may have done this many times when you were outside in the cold without gloves. The friction between your hands converts some of the kinetic energy of the motion into heat. The filament of the bulb does essentially the same thing, converting electrical energy into heat and light.

In our water analogy, resistance might be thought of as a partial blockage in a pipe that decreases the rate of the flow of the water. In terms of electricity, the resistance of a material is defined as the ratio of the potential difference across the material to the current through the material.

$$\text{resistance} = \frac{\text{potential difference}}{\text{current}} \quad \text{or} \quad R = \frac{\Delta V}{I}$$

This relationship is known as Ohm's law, and it is one of the most important formulas you need to use when you study electricity in physics. The resistance of a material is measured in ohms (Ω), which are equivalent to volts/amps.

Example 1

Find the current through a bulb that has a resistance of 3.0 Ω and a potential difference of 9.0 V across it.

Given: $\Delta V = 9.0\ V$ $R = 3.0\ \Omega$

Find: I

Isolate:

Starting with: $R = \frac{\Delta V}{I}$

Multiply both sides by I: $I \times R = \frac{\Delta V}{\cancel{I}} \times \cancel{I}$

Divide both sides by R: $\frac{I \times \cancel{R}}{\cancel{R}} = \frac{\Delta V}{R}$

We get: $I = \frac{\Delta V}{R}$

Solution: $I = \frac{\Delta V}{R} = \frac{9.0\ V}{3.0\ \Omega} = \frac{9.0\ \cancel{V}}{3.0\ \cancel{V}/A} = \mathbf{3.0\ A}$

Ohm's law is not really a universal or fundamental law, as not all materials "obey" it. Materials that follow Ohm's law over a wide range of potential differences are called **ohmic materials**. Materials such as semiconductors, which don't follow Ohm's law over a wide range of potential differences, are called **nonohmic**.

In reality, the copper wire in our flashlight actually offers a small amount of resistance to the charges. The resistance is relatively low, so the wire is considered a **conductor**. A conductor is a material that allows charges to move freely through it. In addition to conductors and resistors, there are **semiconductors**, which conduct electricity better than resistors, but not as well as conductors.

Four factors affect the resistance of a component:

1. The **material** the component is made of. For example, copper is a good conductor, offering less resistance than a material such as tungsten.
2. The **temperature** of the component. The molecules of a hot material exhibit more motion, leading to more collisions with charges, resulting in greater resistance.
3. The **cross-sectional area** of the component. Just as a wider pipe allows for a greater flow of water, a wider wire offers less resistance, allowing for a greater flow of charge.
4. The **length** of the component. For example, using a very long extension cord for an electric hedge-trimmer results in greater resistance, decreased current, and decreased power.

Resistivity

At a given temperature, the resistance offered by a material will be proportional to its length and inversely proportional to its cross-sectional area.

$$R \propto \frac{L}{A}$$

If we insert the proportionality constant for a given material into the formula, we can set the sides equal to each other.

$$R = \rho \frac{L}{A}$$

This quantity (ρ) is not a true constant, because it varies from material to material and from temperature to temperature. We call this quantity *resistivity* (ρ), and it is measured in $\Omega \cdot m$.

Example 2

A wire with a length of 0.895 m and a cross-sectional area of 1.0×10^{-7} m^2 has a current of 1.00×10^2 A through it and a potential difference of 9.0 V across it. Find the resistivity of the wire.

Given: $L = 0.895$ m $A = 1.0 \times 10^{-7}$ m^2 $I = 1.00 \times 10^2$ A
$\Delta V = 9.0$ V

Find: ρ

As you can see, we don't have quite enough "givens" to solve for resistivity using only our resistivity formula, as we are missing both ρ and R. We can use Ohm's law to find R, because we have also been given I and ΔV.

$$R = \frac{\Delta V}{I} = \frac{9.0\ \text{V}}{1.00 \times 10^{2}\ \text{A}} = \mathbf{0.090\ \Omega}$$

Isolate:

Starting with: $R = \rho \frac{L}{A}$

Multiply both sides by $\frac{A}{L}$: $R \times \frac{A}{L} = \rho \frac{\not{L}}{\not{A}} \times \frac{\not{A}}{\not{L}}$

We get: $\rho = R\frac{A}{L} = \frac{(0.090\ \Omega)(1.0 \times 10^{-7}\ \text{m}^2)}{0.895\ \text{m}} = \mathbf{1.0 \times 10^{-8}\ \Omega \cdot m}$

Resistors

We said that bulbs could be considered resistors because they resist the flow of charge through them. They are not, however, the only type of resistors. There are also components whose sole purpose is to add resistance to a circuit. Adding resistance to the circuit will decrease the current and protect other components from overloading.

Lesson 6–2 Review

1. ______________ is the opposition to the flow of electric current.
2. A component has a potential difference of 3.0 V across it and a current of 2.0 A through it. Calculate its resistance.
3. Find the current through a 3.0 Ω resistor with a voltage of 1.5 V across it.

Lesson 6–3: Electric Power

We said that the bulb in our flashlight glows because the filament in the bulb is a resistor, which converts the electrical energy of the charged particles into heat and light. However, you may have noticed that the bulbs that you buy are sorted by watts (W), a unit of power—not ohms, a unit of resistance. We encountered the unit called watts when we studied power in Chapter 3. In this lesson we will study **electric power**, which is the rate at which electrical energy is converted into other forms of energy, such as heat, light, and/or kinetic energy.

Formulas for Electric Power

$P = I\Delta V$ $\quad P = I^2R$ $\quad P = \frac{\Delta V^2}{R}$ $\quad P = \frac{W}{\Delta t}$

Example 1

A bulb operates at 40.0 mW when connected to a 9.0 V battery. Find both the current through and the resistance offered by the bulb.

Convert: The small "m" before the symbol for watts (as in 40.0 mW) stands for "milli" (1 W = 1000 mW).

$$40.0\ \cancel{mW} \times \frac{1\ W}{1000\ \cancel{mW}} = \mathbf{4.00 \times 10^{-2}\ W}$$

Notice that the way we write the answer, 4.00×10^{-2} W, retains the same number of significant digits as our original value of 40.0 mW.

Given: $P = 4.00 \times 10^{-2}$ W $\quad \Delta V = 9.0$ V

Find: I and R

Isolate:

Starting with: $P = I\Delta V$

Divide both sides by ΔV: $\frac{P}{\Delta V} = \frac{I\cancel{\Delta V}}{\cancel{\Delta V}}$

We get: $I = \frac{P}{\Delta V}$

Solution: $I = \frac{P}{\Delta V} = \frac{4.00 \times 10^{-2}\ W}{9.0\ V} = \mathbf{0.0044\ A}$ or $\mathbf{4.4\ mA}$

Let's not use this value for I to find R. Remember, I always recommend using only the original givens to find unknowns in multiple part questions (whenever possible) to avoid carrying over errors to other calculations.

Isolate:

Starting with: $P = \frac{\Delta V^2}{R}$

Multiply both sides by R: $P \times R = \frac{\Delta V^2}{\cancel{R}} \times \cancel{R}$

Divide both sides by P: $\frac{\cancel{P}R}{\cancel{P}} = \frac{\Delta V^2}{P}$

We Get: $R = \frac{\Delta V^2}{P}$

Solution: $R = \frac{\Delta V^2}{P} = \frac{(9.0\ V)^2}{4.00 \times 10^{-2}\ W} = 2025\ \Omega$

$= \mathbf{2.0 \times 10^3\ \Omega}$ (after rounding)

The Cost of Electric Power

Many textbooks stress the fact that what we call "power companies" really charge us for energy, not power. The "power" company doesn't care about the rate that we use the electrical energy, just the total amount that we use. If we use 2500 joules in 10 minutes or in 10 days, it still costs us the same amount. If you ever look at the bill that comes from you "power" company, you will see that you are being charged a certain amount of money per kilowatt-hour (kW-h). Kilowatts are units of power, and hours are units of time. If you multiply power by time, you get work, which can be measured in joules, or kilowatt-hours.

$$P \times T = \frac{W}{\cancel{T}} \times \cancel{T} = W$$

Lightbulbs and many of your appliances list their power ratings somewhere on them. Other appliances will list the current they draw and the operating voltage. Check on the back of your TV or the bottom of your toaster oven. By checking the power rating of various appliances and the amount of money your family is being charge per kilowatt-hour, you can calculate approximately how much it is costing you (or your parents) to run these appliances.

The Cost of Electric Power

$$\text{Cost} = \text{Energy} \times \text{Rate}$$
$$C = ER$$

Example 2

My toaster oven has a power rating of 1500 W. If I am charged \$0.12 per kilowatt-hour, and I run my toaster oven for 8.0 minutes every day, approximately how much is this costing me per month?

Convert: We must change watts to kilowatts.

$$1500\,\cancel{\text{W}} \times \frac{1\,\text{kW}}{1000\,\cancel{\text{W}}} = \mathbf{1.5\,kW}$$

Let's also convert 8.0 minutes per day to hours/month.

$$\frac{8.0\,\cancel{\text{min}}}{\cancel{\text{day}}} \times \frac{1\,\text{hour}}{60\,\cancel{\text{min}}} \times \frac{30\,\cancel{\text{day}}}{\text{month}} = \mathbf{4.0\,hours/month}$$

Now, we can calculate the number of kW-h I am using to run the toaster each month.

$$1.5\,\text{kW} \times 4.0\,\text{h} = \mathbf{6.0}\,\text{kW-h}$$

Given: Energy (E) = 6.0 kW-h Rate (R) = \$0.12/kW-h

Find: Cost (C)

Solution: $C = ER = (6.0\,\cancel{\text{kW-h}})(\$\,0.12/\cancel{\text{kW-h}}) = \mathbf{\$0.72}$

So, I am only spending about 72 cents to run my toaster each month. It is always surprising to see how little certain appliances cost to use. It may make you wonder why the electric bill can get so high at times, or why your parents always told you to shut off the lights when you left the room. Remember, there are probably quite a few appliances and bulbs in your home, and all of those little charges add up. More importantly, some appliances with higher power ratings probably account for a relatively large portion of your monthly bill. Refrigerators, air conditioners, dryers, and hot-water heaters draw much more power than your toaster oven, so they cost your family more per hour that they are in operation.

Lesson 6–3 Review

1. What is the current through a 40.0 W bulb with a voltage of 120 V across it?
2. A hair dryer draws 8.2 A of current from a 120-volt outlet. What is the power rating for the toaster?
3. Calculate the current through a 3.0 Ω resistor that operates at 60.0 W.

Lesson 6–4: Circuits and Schematic Diagrams

Two major skills in this area of physics involve being able to draw and being able to interpret schematic diagrams for circuits. **Schematic diagrams** are visual representations of circuits, using standard symbols to represent individual components. Another skill involves being able to construct some of the circuits that you create diagrams for, but because that involves materials that don't come with this book, I can't help you with that. Hopefully, you will find opportunities in lab or at home to construct some simple circuits, such as the homemade flashlight we have been discussing in this chapter.

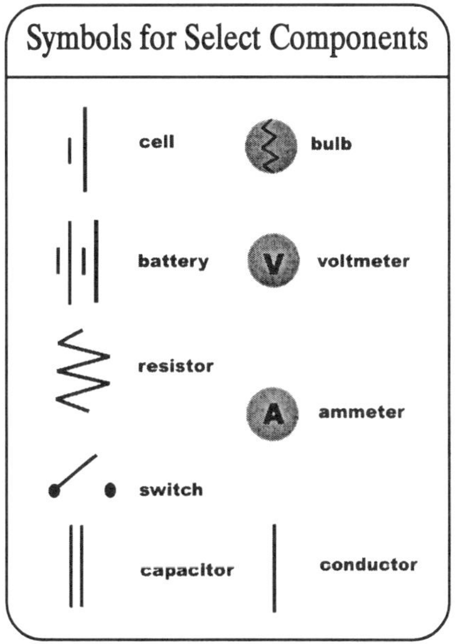

Series Circuit

There are two main types of circuits for which we will discuss and create diagrams. A **series circuit** is a circuit in which there is only one pathway for the charges to travel through. The components in a series circuit are arranged end to end.

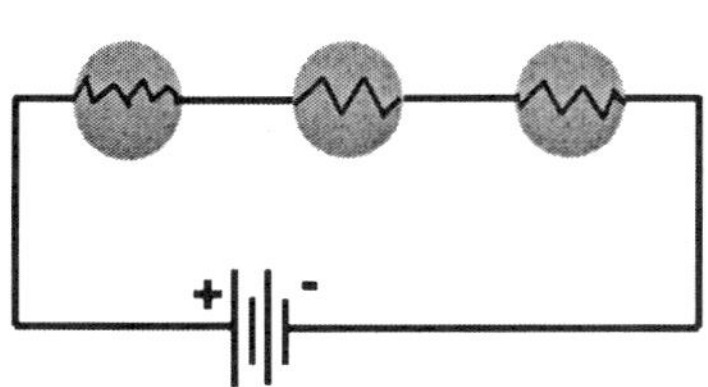

Figure 6.2

Because there is only one pathway for the flow of electricity in a series circuit, the current measured at any point in the circuit will be the same. Therefore, the total current in a series circuit (I_s) is equal to the current through each of the components. So, the current through each of the bulbs in Figure 6.2 will be the same.

$$I_S = I_1 = I_2 = \ldots I_x$$

Allow me to explain the notation that I used in this, and each of the following formulas. When I write "$\ldots I_x$," I mean that you follow this pattern for as many components as you have in your circuit. If you only have two, you would stop at I_2, but if you have five, you would continue with the pattern until I_5.

Each of the bulbs in the picture for a series circuit act as a resistors, so the more you add, the greater the total resistance. If you like the water analogy for electric current, each resistor is like a partial blockage in a water pipe. The effect of numerous blockages is cumulative, resulting in an increase in resistance and a decrease in current. So, the total resistance of the individual resistors in series is equivalent to using one resistor with a resistance of R_s.

$$R_S = R_1 + R_2 + \ldots R_x$$

When components, such as our bulbs, are connected in series, the potential difference across the terminals of the battery is equal to the sum of all the potential differences or "voltage drops" across the components.

$$V_S = V_1 + V_2 + \ldots V_x$$

There aren't any capacitors shown in our diagram of a series circuit (Figure 6.2), but when more than one capacitor is connected in a series, the total capacitance, which is less than the smallest capacitance in the group, can be determined with the following formula:

$$\frac{1}{C_s} = \frac{1}{C_1} + \frac{1}{C_2} + \ldots \frac{1}{C_x}$$

Parallel Circuits

In a **parallel circuit** there is more than one pathway for the electricity to travel through. The components in a parallel circuit are arranged side by side, and they can each trace a direct path back to the source of potential difference.

Because each of the bulbs can trace a path back to each side of the battery, the potential difference across each bulb is the same as the potential difference supplied by the battery.

$$V_p = V_1 = V_2 = \ldots V_x$$

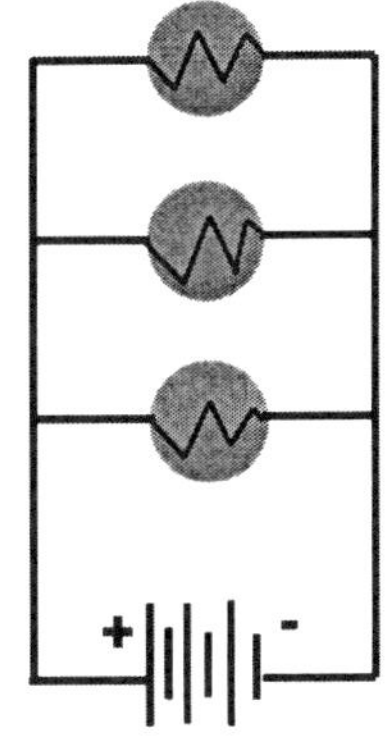

Figure 6.3

You might think that having more resistors in parallel would lead to more resistance, but, remember, in addition to adding resistors, you are also adding other pathways for the current to flow through. Using the water analogy once again, more water may be able to flow through three pipes, each with a partial blockage, than can flow through one clear pipe. You will find that the total (or equivalent) resistance in a parallel circuit will always be lower than any of the individual resistances, as shown by the formula here:

$$\frac{1}{R_p} = \frac{1}{R_1} + \frac{1}{R_2} + \ldots \frac{1}{R_x}.$$

The current in each of the parallel pathways need not be the same. If the individual bulbs in our circuit each provided different resistances, for example, there would be different amounts of current flowing through each path. You may have heard the expression about "following the path of least resistance." The current is greatest in the part of the circuit that connects to all of the parallel branches, but current is also greater in the branch with lower resistance that a path with greater resistance.

$$I_p = I_1 + I_2 \ldots I_x$$

There are no capacitors shown in our picture of a parallel circuit, but when capacitors are arranged parallel to each other, the equivalent capacitance may be found with the following formula.

$$C_p = C_1 + C_2 + \ldots C_x$$

Comparing Series and Parallel Circuits

Series Circuit	**Parallel Circuit**
$I_S = I_1 = I_2 = \ldots I_x$	$I_p = I_1 + I_2 \ldots I_x$
$R_S = R_1 + R_2 + \ldots R_x$	$\frac{1}{R_p} = \frac{1}{R_1} + \frac{1}{R_2} + \ldots \frac{1}{R_x}$
$V_S = V_1 + V_2 + \ldots V_x$	$V_p = V_1 = V_2 = \ldots V_x$
$\frac{1}{C_s} = \frac{1}{C_1} + \frac{1}{C_2} + \ldots \frac{1}{C_x}$	$C_p = C_1 + C_2 + \ldots C_x$

Before we get to the harder types of problems involving many calculations, let's try some easy examples first, just to get use to working with the individual formulas.

Example 1

A 3.0 Ω resistor is connected in parallel to a 5.0 Ω resistor. What is their equivalent (total) resistance?

Given: $R_1 = 3.0\ \Omega$ $R_2 = 5.0\ \Omega$

Find: R_p

Solution: $\frac{1}{R_p} = \frac{1}{R_1} + \frac{1}{R_2} = \frac{1}{3.0\ \Omega} + \frac{1}{5.0\ \Omega} = 1.875\ \Omega = \mathbf{1.9\ \Omega}$

Note, if you got an answer that looks like 0.53333, then you still need to find the inverse of that value in order to get the final answer, $(0.53333)^{-1} = 1.875$, which we round to 1.9.

Example 2

Three resistors, with resistances of 2.0 Ω, 4.0 Ω, and 6.0 Ω are connected in a series to a 3.0 V battery. Find the equivalent resistance and the current in the circuit.

Given: $R_1 = 2.0\ \Omega$ $R_2 = 4.0\ \Omega$ $R_3 = 6.0\ \Omega$ $V_s = 3.0\ \text{V}$

Find: R_s and I_s

Solution: $R_S = R_1 + R_2 + R_3 = 2.0\ \Omega + 4.0\ \Omega + 6.0\ \Omega = \mathbf{12.0\ \Omega}$

$$I_s = \frac{V_s}{R_s} = \frac{3.0\ \text{V}}{12.0\ \Omega} = \mathbf{0.25\ A}$$

Do you recognize the second formula? It is Ohm's law from Lesson 6–1.

Example 3

A 4.0 Ω and a 6.0 Ω resistor are connected in a series with a 3.0 V battery. Find the voltage across each resistor.

Given: $R_1 = 4.0\ \Omega$ $R_2 = 6.0\ \Omega$ $V_s = 3.0\ \text{V}$

Find: V_1 and V_2

Solution: First, let's find the equivalent resistance of the two resistors.

$$R_S = R_1 + R_2 = 4.0\,\Omega + 6.0\,\Omega = \mathbf{10.0\,\Omega}$$

Next, let's use the voltage and the equivalent resistance to find the current in the circuit.

$$I_s = \frac{V_s}{R_s} = \frac{3.0\text{ V}}{10.0\,\Omega} = \mathbf{0.30\text{ A}}$$

Now, we can find the voltage across each resistor by using Ohm's law. Make sure that you use the resistance of the individual resistors, not the total.

$$V_1 = I_s R_1 = (0.30\text{ A})(4.0\,\Omega) = \mathbf{1.2\text{ V}}$$

$$V_2 = I_s R_2 = (0.30\text{ A})(6.0\,\Omega) = \mathbf{1.8\text{ V}}$$

Notice that the voltage across each of the individual resistors will add up to (1.2 V + 1.8 V = 3.0V) the voltage of the battery. That is a good way to check you answer.

Now, let's go over some examples of interpreting circuit diagrams together. There are many calculations involved in these types of problems, so we will summarize our results in a table at the end. Your instructor may ask you to do the same.

Example 4

Three resistors, with resistances of 3.0 Ω, 6.0 Ω, and 9.0 Ω, respectively, are connected in a series and attached to a 9.0 V battery, as shown in Figure 6.4. Find the total resistance (R_s), the total current (I_s), the current through each resistor, the voltage across each resistor, and the power converted by each resistor.

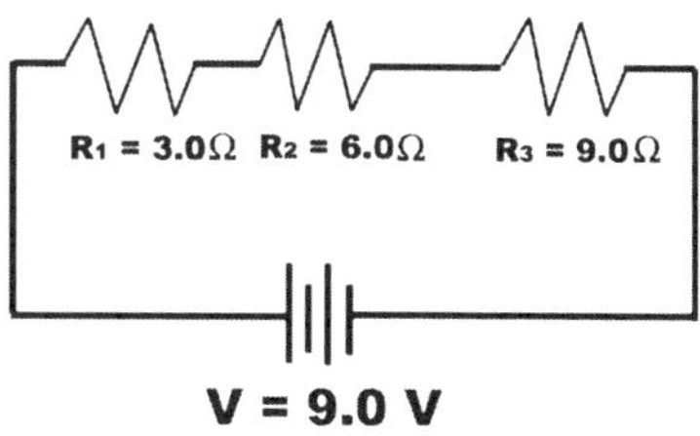

Figure 6.4

To help organize our calculations and answers, let's make a chart that shows all of the required and given information. As you find an answer, you can fill in the appropriate spot in the table.

	R	I	V	P
Resistor 1	$R_1 = 3.0\ \Omega$			
Resistor 2	$R_2 = 6.0\ \Omega$			
Resistor 3	$R_3 = 9.0\ \Omega$			
Totals	R_s	I_s	$V_s = 9.0$ V	P_s

In a series circuit, the logical place to start is to find the total (equivalent) resistance in the circuit.

$$R_s = R_1 + R_2 + R_3 = 3.0\,\Omega + 6.0\,\Omega + 9.0\,\Omega = \mathbf{18.0\,\Omega}$$

Then, we can use Ohm's law to find the total current in the circuit.

$$I_s = \frac{V_s}{R_s} = \frac{9.0\text{V}}{18.0} = \mathbf{0.50\ A}$$

Remember, in a series circuit, current is the same at any point in the circuit, so we can fill in the current through each resistor.

$$I_S = I_1 = I_2 = I_3 = \mathbf{0.50\ A}$$

Now that we know the current and resistance through each resistor, we can find the potential difference across each resistor. Remember to use the individual resistances for these calculations, and check that the total of each potential difference is the same (within rounding) to the voltage of the battery.

$$V_1 = I_1R_1 = (0.50\text{ A})(3.0\,\Omega) = \mathbf{1.5\ V}$$

$$V_2 = I_2R_2 = (0.50\text{ A})(6.0\,\Omega) = \mathbf{3.0\ V}$$

$$V_3 = I_3R_3 = (0.50\text{ A})(9.0\,\Omega) = \mathbf{4.5\ V}$$

The sum of our individual potential drops (1.5 V + 3.0 V + 4.5 V = 9.0 V) is equal to the voltage of our battery, so we can feel confident that we are still on track.

Finally, we need to calculate the power dissipated by each of the resistors. We can choose from a couple of the formulas for electric power that we went over in Lesson 6–3. I will work with the formula P = IV. The current through each resistor is the same, but make sure you use the individual resistances when calculating the power for each resistor.

$$P_1 = I_1V_1 = (0.50\text{ A})(1.5\text{ V}) = \mathbf{0.75\ W}$$

$$P_2 = I_2V_2 = (0.50\text{ A})(3.0\text{ V}) = \mathbf{1.5\ W}$$

$$P_3 = I_3V_3 = (0.50\text{ A})(4.5\text{ V}) = \mathbf{2.25\ W}\ ^*$$

(* Technically, this value should be rounded to 2.3 W, but this will result in slight differences when I compare the total power.)

Now we can check these answers by determining the total power for the circuit with two different methods.

$$P_s = I_sV_s = (0.5\text{ A})(9.0\text{ V}) = \mathbf{4.5\ W}$$

$$P_s = P_1 + P_2 + P_3 = 0.75\text{ W} + 1.5\text{ W} + 2.25\text{ W} = \mathbf{4.5\ W}$$

The fact that we ended up with the same answer for the total power with both calculations leads me to believe that our calculations are correct. Now, let's add the values that we have found to our chart.

	R	I	V	P
Resistor 1	$R_1 = 3.0\ \Omega$	$I_1 = 0.50$ A	$V_1 = 1.5$ V	$P_1 = 0.75$ W
Resistor 2	$R_2 = 6.0\ \Omega$	$I_2 = 0.50$ A	$V_2 = 3.0$ V	$P_2 = 1.5$ W
Resistor 3	$R_3 = 9.0\ \Omega$	$I_3 = 0.50$ A	$V_3 = 4.5$ V	$P_3 = 2.25$ W
Totals	$R_s = 18.0\ \Omega$	$I_s = 0.50$ A	$V_s = 9.0$ V	$P_s = 4.5$ W

Now, let's try a similar problem dealing with a parallel circuit.

Example 5

Three resistors, with resistances of 5.0 Ω, 8.0 Ω, and 12.0 Ω, respectively, are connected in a parallel circuit with a 9.0 V battery, as shown in Figure 6.5. Find the total resistance (R_s), the total current (I_s), the current through each resistor, the voltage across each resistor, and the power converted by each resistor.

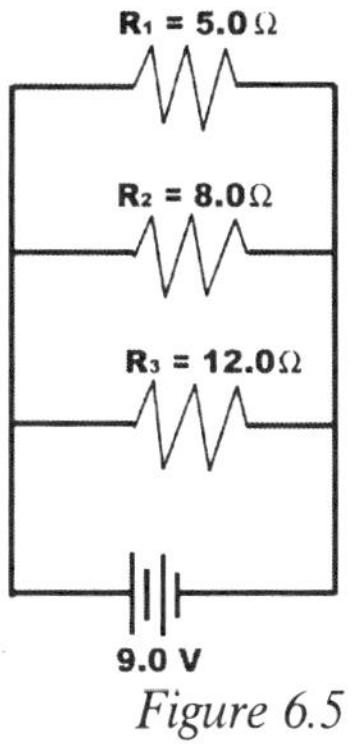

Figure 6.5

The first thing that I think of when I do a parallel problem such as this is that the voltage across each component will be the same as the voltage of the source, because each component can trace a path back to each terminal of the battery.

$$V_p = V_1 = V_2 = V_3 = \mathbf{9.0\ V}$$

Once again, let's start a table that will help us organize all of the information that we need to provide to answer this question. We will start with what we already know.

	R	I	V	P
Resistor 1	$R_1 = 5.0\,\Omega$	$I_1 =$	$V_1 = 9.0\ V$	$P_1 =$
Resistor 2	$R_2 = 8.0\,\Omega$	$I_2 =$	$V_2 = 9.0\ V$	$P_2 =$
Resistor 3	$R_3 = 12.0\,\Omega$	$I_3 =$	$V_3 = 9.0\ V$	$P_3 =$
Totals	R_p	$I_p =$	$V_p = 9.0\ V$	$P_p =$

I would begin by finding the equivalent resistance for the three resistors, as shown here:

$$\frac{1}{R_p} = \frac{1}{R_1} + \frac{1}{R_2} + \frac{1}{R_3} = \frac{1}{5.0\,\Omega} + \frac{1}{8.0\,\Omega} + \frac{1}{12.0\,\Omega} = \mathbf{2.4\,\Omega}$$

Next, let's use Ohm's law to find the current through the individual resistors.

$$I_1 = \frac{V_1}{R_1} = \frac{9.0\ V}{5.0\,\Omega} = \mathbf{1.8\ A}$$

$$I_2 = \frac{V_2}{R_2} = \frac{9.0\ V}{8.0\,\Omega} = 1.125 = \mathbf{1.1\ A}$$

$$I_3 = \frac{V_3}{R_3} = \frac{9.0\ V}{12.0\,\Omega} = \mathbf{0.75\ A}$$

Let's find the total current in the circuit using two different methods to check our answers.

$$I_p = I_1 + I_2 + I_3 = 1.8\ A + 1.1\ A + 0.75\ A = \mathbf{3.7\ A}$$

$$I_p = \frac{V_p}{R_p} = \frac{9.0\ V}{2.4\,\Omega} = 3.75\ A = \mathbf{3.8\ A}$$

Although not identical, our answers are close enough to confirm that we did the calculations correctly. The differences can be accounted for by the rounding that we have done at each step.

Now we can use the electric power formula to find the power dissipated by each resistor.

$$P_1 = I_1V_1 = (1.8)(9.0\text{ V}) = 16.2\text{ W} = \mathbf{16\text{ W}}$$

$$P_2 = I_2V_2 = (1.1)(9.0\text{ V}) = \mathbf{9.9\text{ W}}$$

$$P_3 = I_3V_3 = (0.75)(9.0\text{ V}) = 6.75\text{ W} = \mathbf{6.8\text{ W}}$$

Let's find the total power with two different methods to check our calculations.

$$P_p = P_1 + P_2 + P_3 = 16\text{ W} + 9.9\text{ W} + 6.8\text{ W} = 32.7\text{ W} = \mathbf{33\text{ W}}$$

$$P_p = I_pV_p = (3.8\text{ A})(9.0\text{ V}) = \mathbf{34\text{ W}}$$

Once again, the slight difference in our answers can be attributed to rounding.

Finally, let's summarize the results of all of our calculations by filling in our table, as shown here:

	R	I	V	P
Resistor 1	$R_1 = 5.0\ \Omega$	$I_1 = 1.8$ A	$V_1 = 9.0$ V	$P_1 = 16$ W
Resistor 2	$R_2 = 8.0\ \Omega$	$I_2 = 1.1$ A	$V_2 = 9.0$ V	$P_2 = 9.9$ W
Resistor 3	$R_3 = 12.0\ \Omega$	$I_3 = 0.75$ A	$V_3 = 9.0$ V	$P_3 = 6.8$ W
Totals	$R_p = 2.4\ \Omega$	$I_p = 3.7$ A	$V_p = 9.0$ V	$P_p = 33$ W

In addition to interpreting circuit diagrams, you will want to be able to draw and perhaps even construct them based on specific instructions. A circuit diagram isn't like a calculation in that there is a specific answer. There is room for some difference between two different answers. However, certain features are required for each diagram in order to be considered correct. Let's try an example.

Example 6

Construct a schematic diagram of a battery, a switch, and three bulbs connected in a parallel circuit. Arrange the components in such a way that the switch controls all three lights at once.

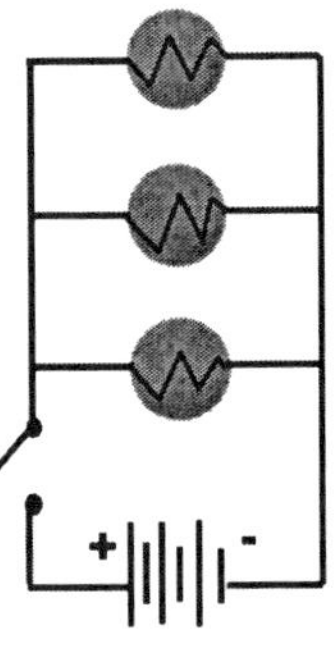

Figure 6.6

Solution:

In order to have the switch interrupt the flow of electricity to all three bulbs when it is open, we must put the switch on a pathway between the battery and all three bulbs.

Of course, there is room for some differences between individual answers when you are asked to draw schematic diagrams. For example, you may have placed your switch on the left-hand side, rather than the right. There are some aspects of the diagram that must be specific, however. Notice in our diagram, for example, that when the switch is open, none of the bulbs can trace a path back to the right side of the battery.

Before we move on to our review, I will take this opportunity to explain the difference between two of the symbols that appeared in our table of schematic symbols. Many of the components that we normally call *batteries*, with names such as AAA, D, or C, are actually considered *cells*. If you look on a AAA cell, you will note that it provides 1.5 V of potential difference. A battery is actually a group of cells arranged in series. If you dissect a 9 V battery, you will find that underneath its metallic coating are six, tiny 1.5 V cells. The 9 V (6×1.5 V = 9.0 V!) is the only "true" battery out of the tiny components that most of us call batteries.

Lesson 6–4 Review

1. Three resistors with resistances of 2.0 Ω, 5.0 Ω, and 9.0 Ω, respectively, are connected in a series circuit to a 9.0 V battery. What is the current through the 5.0 Ω resistor?

2. A 2.0 μf capacitor is connected in a parallel circuit to a 7.0 μf capacitor. What is the equivalent capacitance?

3. Three resistors with resistances of 12.0 Ω, 14.0 Ω, and 22.0 Ω, respectively, are connected in a parallel circuit to a 12.0 V battery. Find the power dissipated by the 14.0 Ω resistor.

Chapter 6 Examination

Part I—Matching

Match the following terms to the definitions that follow.

a. current
b. circuit
c. resistance
d. conventional current
e. alternating current
f. direct current
g. electric power
h. Ohm's Law
i. resistivity

_____1. Electric current that periodically reverses directions.

_____2. Visualization of current flowing out of the positive terminal into the negative terminal.

_____3. An arrangement of components that provides one or more pathways for electric current to flow.

_____4. The rate at which electrical energy is transformed.

_____5. The opposition to the flow of electric current.

Part II—Multiple Choice

For each of the following questions choose the best answer.

6. What units are used to measure resistance?
   a) amps b) volts c) ohms d) watts

7. What units are used to measure power?
   a) amps b) volts c) ohms d) watts

8. Which of the following quantities are measured in units called *volts*?
   a) current
   b) potential difference
   c) potential energy
   d) work

9. A wire carries a current of 3.0 A. How much charge passes by a point in 90.0 seconds?
   a) 270 C b) 30 C c) 87 C d) 93 C

10. Calculate the current through a 3.0 Ω resistor with a voltage of 9.0 V across it.
    a) 0.33 A b) 27 A c) 3.0 A d) 6.0 A

11. Three resistors with resistances of 2.0 Ω, 4.0 Ω, and 7.0 Ω, respectively, are connected in a series circuit. Find the equivalent resistance.
    a) 1.1 Ω b) 13.0 Ω c) 56.0 Ω d) 93 Ω
12. Three resistors with resistances of 1.0 Ω, 7.0 Ω, and 9.0 Ω, respectively, are connected in a parallel circuit to a 9.0 V battery. Find the potential difference across the 9.0 Ω resistor.
    a) 1.0 V b) 1.5 V c) 9.0 V d) 12.0 V
13. Find the power dissipated by a 7.0 Ω resistor with a current of 2.0 A through it.
    a) 5.0 W b) 9.0 W c) 14 W d) 28 W
14. What is the power dissipated by a 5.0 Ω resistor with a potential difference of 9.0 V across it?
    a) 4.0 W b) 5.0 W d) 14 W d) 16 W

Part III—Calculations

Perform each of the following calculations.

15. 120 C of charge pass a point in a wire every minute. What is the current in the wire?
16. A bulb has a potential difference of 6.0 V across it and a current of 1.0 A through it. Calculate its resistance.
17. Calculate the current through a 9.0 Ω resistor that operates at 30.0 W.
18. What is the power developed in a 6.0 Ω resistor with a potential of 1.5 V across it?
19. Three resistors with resistances of 5.0 Ω, 9.0 Ω, and 12.0 Ω, respectively, are connected in a series circuit with a 6.0 V battery. Find the potential difference across the 5.0 Ω resistor.
20. My electric fan draws 0.55 A of current with a potential difference of 120 V across it. If I pay $0.17/kW-h, how much will it cost me to operate this fan for 3.00 hours?

Answer Key

The actual answers will be shown in brackets, followed by an explanation. If you don't understand an explanation that is given in this section, you may want to go back and review the lesson that the question came from.

Lesson 6–1 Review

1. [direct current]—Batteries and cells, like the ones that we use in flashlights, provide direct current.

2. [110 C]—$q = I\Delta t = (2.0\ A)(55.0\ s) = 110\ C$

3. [5.0 A]—$I = \frac{q}{\Delta t} = \frac{175\ C}{35\ s} = 5.0\ A$

Lesson 6–2 Review

1. [resistance]—Given the potential difference and current, we can calculate resistance with Ohm's law: $R = \frac{\Delta V}{I}$.

2. [1.5 Ω]—$R = \frac{\Delta V}{I} = \frac{3.0\ V}{2.0\ A} = 1.5\ \Omega$

3. [0.50 A]—$I = \frac{\Delta V}{R} = \frac{1.5\ V}{3.0\ \Omega} = 0.50\ A$

Lesson 6–3 Review

1. [0.33 A]—$I = \frac{P}{\Delta V} = \frac{40.0\ W}{120\ V} = 0.33\ A$

2. [980 W]—$P = I\Delta V = (8.2\ A)(120\ V) = 980\ W$

3. [4.5 A]—$I = \sqrt{\frac{P}{R}} = \sqrt{\frac{60.0\ W}{3.0\ \Omega}} = 4.5\ A$

Lesson 6–4 Review

1. [0.56 A]—You should recall that in a series circuit, there is only one value for current, as shown in the formula $I_S = I_1 = I_2 = \ldots I_x$. If we find the total current, that will be equal to the current through the 5.0 Ω resistor. First, we will find the total resistance.

   $R_S = R_1 + R_2 + R_3 = 2.0\ \Omega + 5.0\ \Omega + 9.0\ \Omega = 16.0\ \Omega$

Then, we will use Ohm's law to find the current.

$$I_s = \frac{V_s}{R_s} = \frac{9.0V}{16.0} = 0.56\ A$$

2. [9.0μf]—$C_p = C_1 + C_2 = 2.0\ \mu f + 7.0\ \mu f = 9.0\ \mu f$

3. [10.3 W]—There are several different formulas for electric power, so there are several different ways to go about solving this problem. You might follow the procedure that I demonstrated with the example problem, where I found the total resistance for the circuit, so that I could calculate the current through each branch of the circuit. However, the easiest way to solve this problem is to select the formula for electric power that doesn't require you to know the current through the resistor,

$$P = \frac{\Delta V^2}{R}.$$

To solve this problem, all you need to remember is that in a parallel circuit, the voltage across each resistor is the same as the voltage of the battery. So,

$$P = \frac{\Delta V^2}{R} = \frac{(12.0\ V)^2}{14.0\ \Omega} = 10.3\ W.$$

It should be noted that if you solved this problem with different formulas, there could be a slight difference due to rounding.

Chapter 6 Examination

1. [e. alternating current]—In the United States, household current reverses itself 120 times per second, for a total of 60 cycles/second.
2. [d. conventional current]—Although the electrons in a particular circuit may flow clockwise, the convention current would be in the opposite direction, counter clockwise.
3. [b. circuit]—A complete circuit is required for current to flow.
4. [g. electric power]—A 60-watt bulb converts 60 joules of energy per second.
5. [c. resistance]—Not to be mistaken with *resistivity*, which is a proportionality constant.
6. [c. ohms]—The symbol for ohms is the Greek letter omega (Ω).
7. [d. watts]—The symbol for watts is W.
8. [b. potential difference]—Voltage is another term for potential difference.

9. [a. 270 C]—$q = I\Delta t = (3.0\text{ A})(90.0\text{ s}) = 270\text{ C}$

10. [c. 3.0 A]—$I = \frac{\Delta V}{R} = \frac{9.0\text{ V}}{3.0\,\Omega} = 3.0\text{ A}$

11. [b. 13.0 Ω]—$R_s = R_1 + R_2 + R_3 = 2.0\,\Omega + 4.0\,\Omega + 7.0\,\Omega = 13.0\,\Omega$

12. [c. 9.0 V]—In a parallel circuit, the potential difference across each component is the same as the source.

    $V_p = V_1 = V_2 = V_3$

13. [d. 28 W]—$P = I^2R = (2.0\text{ A})^2(7.0\,\Omega) = 28\text{ W}$

14. [d. 16 W]—$P = \frac{\Delta V^2}{R} = \frac{(9.0\text{ V})^2}{5.0\,\Omega} = 16\text{ W}$

15. [2.0 A]—$I = \frac{q}{\Delta t} = \frac{120\text{ C}}{60\text{ s}} = 2.0\text{ A}$

16. [6.0 Ω]—$R = \frac{\Delta V}{I} = \frac{6.0\text{ V}}{1.0\text{ A}} = 6.0\,\Omega$

17. [1.8 A]—$I = \sqrt{\frac{P}{R}} = \sqrt{\frac{30.0\text{ W}}{9.0\,\Omega}} = 1.8\text{ A}$

18. [0.38 W]—$P = \frac{\Delta V^2}{R} = \frac{(1.5\text{ V})^2}{6.0\,\Omega} = 0.38\text{ W}$

19. [1.2 V]—We can solve this problem with Ohm's law, V = IR, provided we find the total current through the resistor. In order to do that, we need to find the total resistance, and divide that into the total potential difference.

    $R_S = R_1 + R_2 + R_3 = 5.0\,\Omega + 9.0\,\Omega + 12.0\,\Omega = 26.0\,\Omega$

    $I_s = \frac{V_s}{R_s} = \frac{6.0\text{ V}}{26.0\,\Omega} = 0.23\text{ A}$

    Remember that in a series circuit the total current is the same as the current through each of the components, so ($I_S = I_1 = I_2 = I_3 = 0.23\text{ A}$) the current through the 5.0 Ω resistor is 0.23 A. Our final answer can be determined. $V_1 = I_sR_1 = (0.23\text{ A})(5.0\,\Omega) = 1.2\text{ V}$

20. [$0.034, or 3.4 cents]—To solve this problem, you will need the power rating for the fan, which can be calculated as $P = I\Delta V = (0.55\ \text{A})(120\ \text{V}) = 66\ \text{W}$. Next, I will want to convert this to kW.

$$66\ \cancel{\text{W}} \times \frac{1\ \text{kW}}{1000\ \cancel{\text{W}}} = 0.066\ \text{kW}$$

Now, I will calculate how many kilowatt-hours (kW-h) I use when operating this fan for 3.00 hours: 0.066 kW × 3.00 hours = 0.20 kW-h.

Finally, I multiply the number of kilowatt-hours by the rate to find the cost.

C = ER = (0.20 kW-h)($0.17 /kW-h) = $0.034

So, it costs 3.4 cents to run this fan for 3.00 hours.

Magnetism

At the risk of sounding completely old and out of touch, I have to say that I don't think children spend enough time playing with magnets these days. I have fond memories of the hours that I spent as a child, playing with various magnets. Two bar magnets could keep me busy for hours, as I designed and carried out simple activities to test the limits of their mysterious powers. I was fascinated by the fact that if you played with a magnet and a nail for long enough, the nail would start to exhibit magnetic properties of its own. My father once helped me construct a powerful electromagnet. I loved to power it up and pick up a bunch of nails, and then shut off the power and watch most of the nails fall, as much of the magnet's power disappeared with the electric current. I even had this great toy where you would use a magnetic wand to add iron-filing "hair" to a cartoon bald guy. To this day, I still feel like a child again when I have some magnets to play with. Now, before this starts to sound any more like my grandfather's "When I was young we had to whittle all of our toys out of wood!" speech, let's move on to the lessons on magnets.

Lesson 7–1: Magnets and Magnetic Fields

The word *magnet* is derived from the Greek word *magnetite*. Magnetite is a naturally magnetic ore that the Greeks found in an area of Turkey known as Magnesia. I sometimes wonder what it would have been like to see or even carry out demonstrations with these natural magnets, using them to attract small pieces of iron in an age where the advances of technology had not overshadowed the wonder invoked by forces that can act over a distance.

If you have ever played with magnets, then you know that they don't attract all materials. Of the elements found in nature, only iron, nickel, and cobalt are strongly attracted to magnets, and are called **ferromagnetic**. Alloys of these metals are used to create strong bar magnets, like the ones that you may find in a science lab. It is believed that the magnetic properties of these elements are a result of their atomic structures. Magnetic properties are exhibited by charges in motion, and all of the electrons in atoms are charges in motion. Therefore, all moving electrons should act like tiny magnets. Why aren't all atoms magnetic? The magnetic fields produced by individual electrons are very weak, but when many tiny magnetic fields overlap in the proper orientation in areas called **magnetic domains**, they can act as a bigger magnetic field. In most elements, the tiny magnetic fields work against each other as often as they work together. In natural magnets, enough of the tiny fields work together to produce noticeable magnetic properties.

You can induce magnetic properties in a nail because exposing it to a magnetic field will cause its domains to line up in the proper orientation to allow the smaller magnetic fields to overlap and produce a bigger one. Stroking a nail repeatedly with one pole of a bar magnet, moving in the same direction, will speed up this process. Dropping or heating the nail can make it lose its magnetic properties, because the domains can be knocked out of alignment.

Magnetic Poles

If you spent some time playing with bar magnets, you probably made certain discoveries about them fairly quickly. First, you might notice that they are strongest near the ends, or **poles**. If the poles of your bar magnets are labeled "north" and "south," or "+" and "–," then you would quickly notice that the like poles repel each other and the opposite poles attract each other. If you broke one of your magnets, you would find that instead of separating the two poles, you end up with two bar magnets, each with its own north and south poles. If you hung a bar magnet from a string, you would find that it would rotate and orientate itself in the same direction, provided it was not interfered with by other objects.

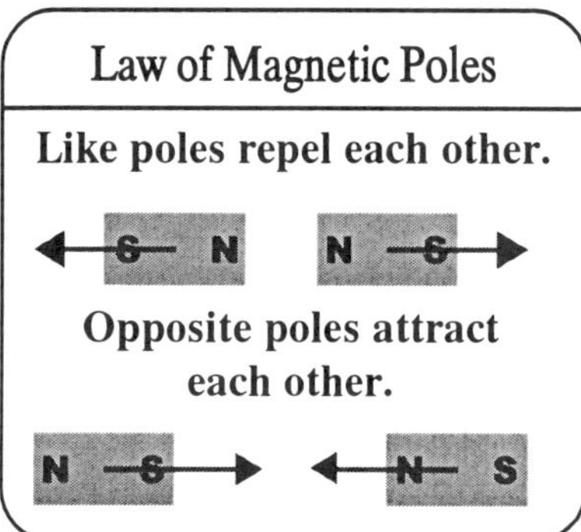
Law of Magnetic Poles
Like poles repel each other.
S N
N S
Opposite poles attract each other.
N S
N S

If you cover a bar magnet with a sheet of paper or thin glass, you can "map" out the magnetic field surrounding the magnet by sprinkling iron filings on the covering. The iron filings line up along the lines of force and give us a two-dimensional picture of a magnetic field.

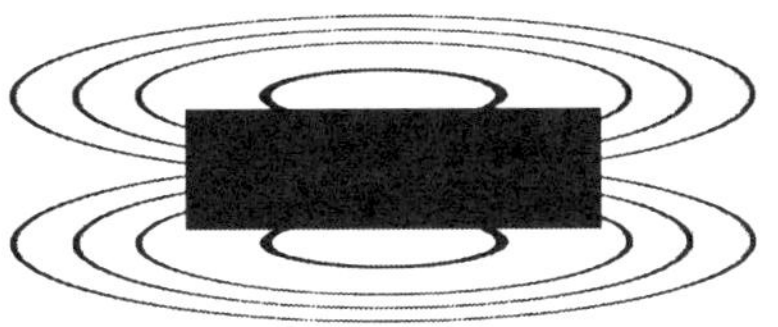

Figure 7.1

Combining this technique with a compass used to determine the direction of the field lines, we could get a clearer picture of the magnetic field around the bar magnet. The differences between the field between two like poles and the field between two unlike poles can be shown with iron filings. The north pole of your compass would point to the south pole of the bar magnet. It is the accepted convention to draw the arrows of the magnetic field diagram originating, or coming out of the north pole of the magnet, and going into the south pole of the magnet.

Earth's Magnetic Field

One potentially confusing aspect concerning the magnetic field of Earth is that the north pole of a compass needle points towards Earth's geographic North Pole. So, the south pole of Earth's magnetic field is actually located near the Earth's geographic North Pole.

When I say that Earth's magnetic south pole is located *near* the geographic North Pole, I mean that the two poles are separated by over 1000 miles. In fact, the actual locations of Earth's magnetic poles change on a daily basis. Following the compass needle directly wouldn't necessarily lead you to Earth's geographic North Pole. Depending on your longitude and latitude, there will be a difference between the lines drawn to the geographic North Pole and the magnetic south pole. We call this difference the **magnetic declination**. Unless you make a correction to your path equal to the magnetic declination, your compass will lead you to Earth's magnetic south pole, rather than its geographic North Pole.

Lesson 7–1 Review

1. A _______________ is a microscopic region in which the magnetic fields of atoms are aligned in the same direction.
2. _______________ materials, such as iron, nickel, and cobalt, are strongly attracted to magnets.

3. Which pole (north or south) of the Earth's magnetic field is located near the Earth's geographic South Pole.

Lesson 7–2: Magnetic Fields Around Current-Carrying Wires

In a story that is often told as if it were the very definition of the word *serendipitous*, Danish physicist Hans Christian Oersted (1777–1851) discovered evidence of the relationship between electricity and magnetism, which would eventually become understood to be a unified force. As the story goes, he completed a simple circuit on his lab table and noticed that it caused the deflection of a nearby compass. With this simple observation, he realized that electric current could affect the motion of magnets. Further studies showed that the compass needle would move when the current was turned on, and again when it was turned off. A current-carrying wire is surrounded by a magnetic field. Electricity produces magnetism.

It is possible to use iron filings and a compass to map out the magnetic field surrounding a current-carrying wire. If you poke a straight, current-carrying wire through a white piece of paper and sprinkle iron filings on the paper, a pattern of concentric circles form around the wire. Using the compass, you could determine whether the magnetic field is clockwise or counterclockwise.

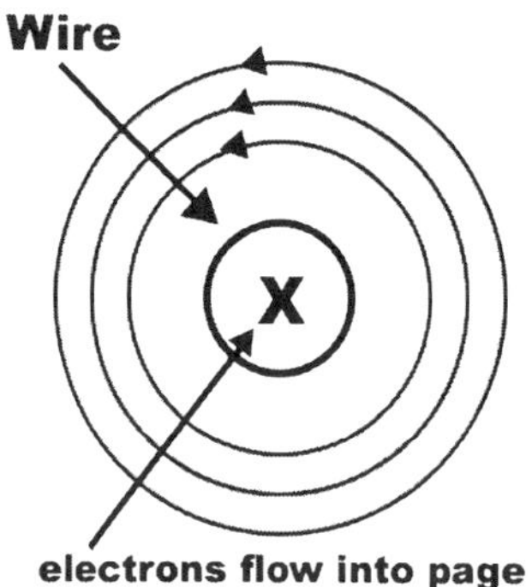

Figure 7.2

Because a compass is not always handy on the day of a test, something called the "right-hand rule" has been developed to determine what the direction of the magnetic field around a current-carrying wire would be.

Before we go over the right-hand rule, let's go over a convention that is used to deal with the fact that, while magnetic fields are three-dimensional, our drawings that are meant to represent magnetic fields are only in two dimensions. We can use arrows to represent directions that are to the left, right, top, or bottom of our page, but we need symbols to show fields or objects that are directed towards us or away from us, meaning "into the page" or "out of the page" we are reading.

Conventions for Vectors

out of the page	into the page
● ● ●	X X X

Figure 7.3

The logic behind the choice of these particular symbols will help you remember which is which. We always use arrows to indicate direction. If an arrow were headed towards you, out of the page, you would see the point of the arrowhead, which might look like our dot symbol. If the arrow were headed away from you, into the page, you would see the back of the feathers, which might look something like our X symbol.

Now, here are the directions for applying the right-hand rule to current-carrying wires.

Steps for Using the Right-Hand Rule to Determine the Direction of the Magnetic Field Around a Current-Carrying Wire

1. Point the thumb of your right hand in the direction of the conventional (+) current. If the picture shows the direction of the electron flow, point your thumb in the opposite direction.
2. Allow the other four fingers of your right hand to curl. They will curl in the direction of the magnetic field around the wire.

Let's try some examples of applying the right-hand rule, which incorporates our two new symbols.

Example 1

Use the right-hand rule to determine the direction of the magnetic field around the current-carrying wire in Figure 7.4.

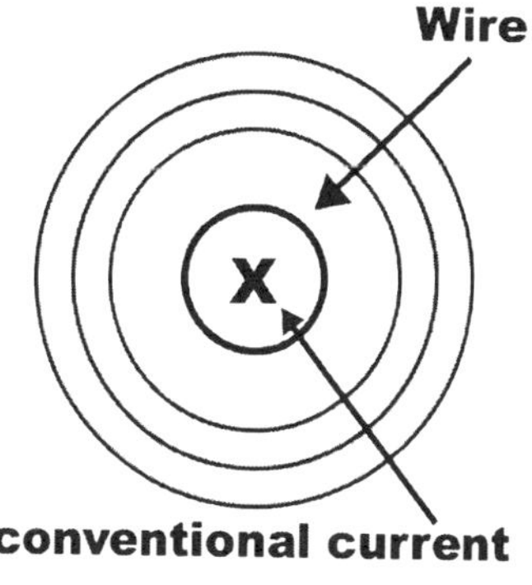

Figure 7.4

Step 1: Point the thumb of your right hand in the direction of the conventional (+) current. If the picture shows the direction of the electron flow, point your thumb in the opposite direction.

The X symbol indicates that the conventional current is directed into the page. Therefore, point your thumb into the page.

Step 2: Allow the other four fingers of your right hand to curl. They will curl in the direction of the magnetic field around the wire.

Your remaining fingers should be curling in a clockwise direction.

Answer: The magnetic field around the wire is in a clockwise direction.

If you were asked to sketch the field on the diagram, it should look like Figure 7.5.

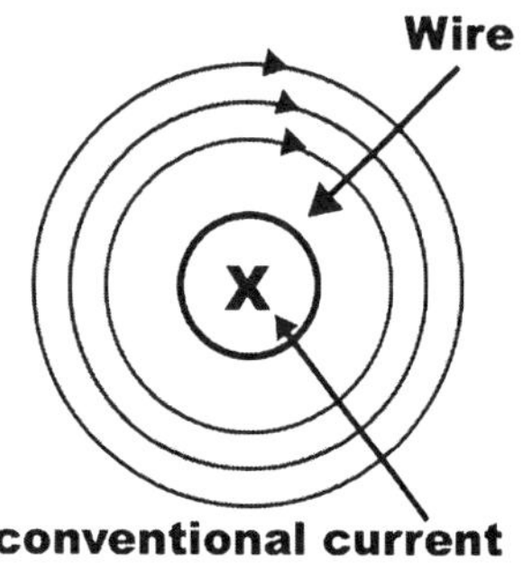

Figure 7.5

Example 2

Use the right-hand rule to determine the direction of the magnetic field around the current-carrying wire in Figure 7.6.

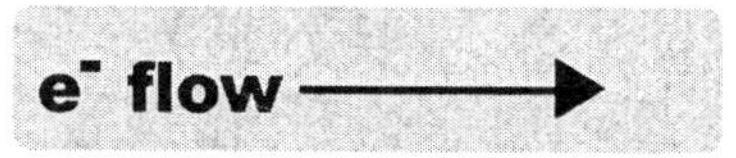

Figure 7.6

Step 1: Point the thumb of your right hand in the direction of the conventional (+) current. If the picture shows the direction of the electron flow, point your thumb in the opposite direction.

The electron (–) flow is to the right, so point your thumb to the left side of the page.

Step 2: Allow the other four fingers of your right hand to curl. They will curl in the direction of the magnetic field around the wire.

Your remaining fingers should be curling towards the page. If you imaging grabbing the wire with your hand, you will see that the magnetic field will be directed into the page above the wire, and out of the page below the wire. This is easier to show with a picture than it is to explain with words.

Answer:

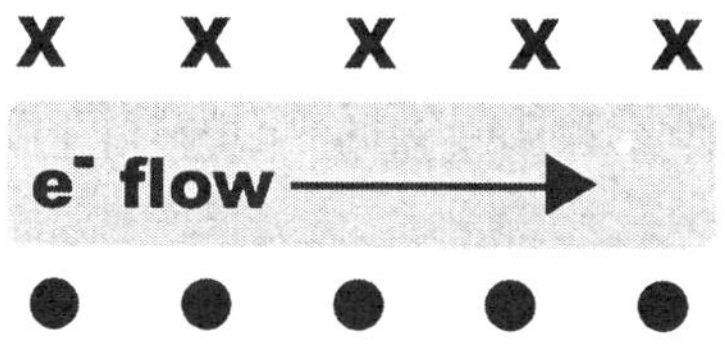

X X X X X

e^- flow →

● ● ● ● ●

Figure 7.7

Magnetic Fields Around Solenoids

If you ever constructed an electromagnet, you probably recall wrapping a wire several times around a piece of metal, such as a nail. When you make a coil of wires such as this, it is called a solenoid. The nail would act as the core of the electromagnet. When current is sent through the solenoid, the magnetic fields of the individual wire loops work together to produce a stronger magnetic field around the solenoid. The solenoid will act like a bar magnet, developing a north pole and a south pole. The right-hand rule can be used to determine the direction of the magnetic field and the identity of the induced poles.

Applying the Right-Hand Rule to Solenoids

1. Figure out the direction of the conventional current through the front of one of the wire loops.
2. Point the thumb of your right hand in the direction of the conventional current.
3. When the other fingers are held out straight, and at right angles to your thumb, they will be pointing towards the south pole of the magnetic field.
4. If you curl the other fingers of your right hand (not your thumb) they will show the direction of the magnetic field through the center of the solenoid and point towards the north pole of the magnetic field.

Lesson 7–2 Review

1. In what direction do the magnetic field lines point inside a solenoid?
2. A wire is orientated "into the page" with the conventional current coming "out of the page," or towards you. What is the direction of the magnetic field (clockwise or counterclockwise) around the wire?
3. Imagine a current-carrying wire with the conventional current directed from left to right. Would the magnetic field be directed into or out of the page directly below (not behind) the wire?

Lesson 7–3: Magnetic Field Strength

Your experiences with magnets have likely shown you that, not only do magnets come in different shapes and sizes, they also vary in a variety of "strengths." A typical bar magnet may only be strong enough to pick up a couple of nails, but the electromagnets used in junkyards are capable of picking up cars. Just as a relatively massive planet such as Jupiter has a stronger gravitational field than one with a smaller mass, such as Mercury, each magnet has its own magnetic field strength. The strength of a magnetic field (B) at a particular point can be measured by a test charge moving through it. The magnitude of the force that a charged particle experiences as it moves perpendicularly through a magnetic field is the product of its charge, velocity, and the strength of the magnetic field.

The Magnitude of the Force Experienced by a Charged Object Moving Perpendicularly Through a Magnetic Field

Force (F) =

charge of particle (q) × velocity of particle (v) × strength of magnetic field (B)

or

$$F = qvB$$

When the velocity of the charged particle is not perpendicular to the direction of the field, we must also include the sine of the angle between the velocity and field strength in our formula.

$$F = qv\sin\theta$$

Because the sine of a 90° angle = 1, we can use the simple formula for dealing with right angles. Note that sin 0° = 0 and sin 180° = 0, so, when a charged particle moves parallel to the magnetic field, it experiences no force.

If we isolate the magnetic field strength (B) in the previous formula, we find that it is proportional to the force and inversely proportional to the charge and velocity of the particle.

$$B = \frac{F}{qv\sin\theta}$$

Magnetic field strength (B) is measured in derived units called tesla (T), where 1 tesla = $1\ N \cdot s^2/C \cdot m = 1 N/A \cdot m$.

Example 1

A proton with a velocity of 1.50×10^6 m/s towards the top of the page experiences a force with a magnitude of 1.80×10^{-13} N as it moves perpendicularly through a magnetic field. Calculate the strength of the magnetic field.

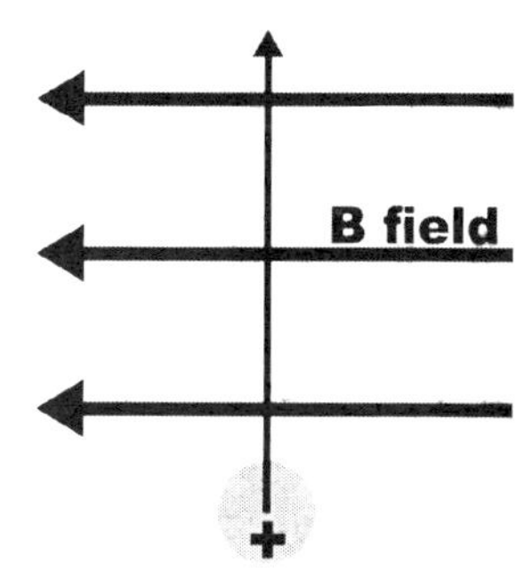

Figure 7.8

Given:

$q = 1.60 \times 10^{-19}$ C $F = 1.80 \times 10^{-13}$ N
$v = 1.50 \times 10^6$ m/s $\theta = 90.0^\circ$

Find: B

Solution:

$$B = \frac{F}{qv\sin\theta} = \frac{1.80 \times 10^{-13}\ \text{N}}{(1.60 \times 10^{-19}\ \text{C})(1.50 \times 10^{6}\ \text{m/s})(\sin 90.0^\circ)} = \mathbf{0.750\ T}$$

Example 2

An oxide ion travels with a velocity of 4.40×10^5 m/s at an angle of 120.0° through a magnetic field with an intensity of 1.50 T. Determine the magnitude of the force acting on the ion.

Given: $q = -3.20 \times 10^{-19}$ C $B = 1.50$ T $v = 4.40 \times 10^5$ m/s
$\theta = 120.0^\circ$

Find: F

Solution:

$$\begin{aligned} F &= qvB\sin\theta \\ &= (3.20 \times 10^{-19}\ \text{C})(4.40 \times 10^{5}\ \text{m/s})(1.50\ \text{T})(\sin 120.0^\circ) \\ &= \mathbf{1.83 \times 10^{-13}\ N} \end{aligned}$$

I removed the negative sign in front of the charge on the ion because the question only asked for the magnitude of the force. If we included the sign for the charge, we would get a negative value for the force, indicating that the direction of the force would be the opposite of the force that the particle would have experienced if it had a positive charge.

Notice that the tesla is equivalent to N/A · m. We know that newtons (N) measure force and amps (A) measure current, but what would the meters (m) refer to? Because charged particles in motion experience a force when moving perpendicularly through a magnetic field, it should come as no surprise that a current-carrying wire will experience a force in a magnetic field, provided the direction of the current is not parallel to the field.

The Magnitude of the Force Experienced by a Current-Carrying Wire in a Magnetic Field

Force (F) = current (I) × length of wire (L) × magnetic field strength (B) × sin of angle between the current and the field.

or

$$F = ILB\sin\theta$$

As with the case of individual charged particles, when the current (I) is perpendicular to the magnetic field (B), we can use a simplified formula.

$$F = ILB$$

When the current is parallel to the magnetic field ($\theta = 0°$), the force will equal zero.

Example 3

A wire segment with a length of 1.50 m has a steady current of 3.50 A through it. What is the magnitude of the force it would experience in a magnetic field measuring 1.20×10^{-2} T if the wire is orientated 20.0° to the magnetic field?

Given: I = 3.50 A L = 1.50 m $B = 1.20 \times 10^{-2}$ T $\theta = 20.0°$

Find: F

Solution:

$$F = ILB\sin\theta = (3.50\text{ A})(1.50\text{ m})(1.20 \times 10^{-2}\text{ T})(\sin 20.0°) = \mathbf{0.0215\text{ N}}$$

Direction of the Force

Now that we know how to calculate the magnitude of the force experienced by charges moving through a magnetic field, let's talk about how to determine the direction in which the force will act. You might assume that the force will be in the direction of the magnetic field, or even in the direction of the particle's velocity, but that is not the case. The direction of the force experience by a charged object is actually perpendicular to both its motion and the direction of the magnetic field.

We use a variation of the right-hand rule, which we will call the "extended right-hand rule" to determine the direction of the force, when we know the directions of both the field and the charges. A variety of questions may be asked to test your knowledge of this extended right-hand rule. Fortunately, these questions are likely to be limited to charges that move perpendicular to the magnetic field.

Applying the Extended Right-Hand Rule to Straight Wires

1. Point the thumb of your right hand in the direction of the velocity (v) of a positive charge. If the picture shows a negative charge, point your thumb in the opposite direction of its motion.
2. Point the index finger of your right hand in the direction of the magnetic field (B).
3. Point the middle finger of your right hand in a direction that is perpendicular to both the thumb and index finger of your right hand. The middle finger gives you the direction of the force (F).

Example 4

Look at Figure 7.9, and use the right-hand rule to determine the direction of the force experienced by a charged object moving through a magnetic field.

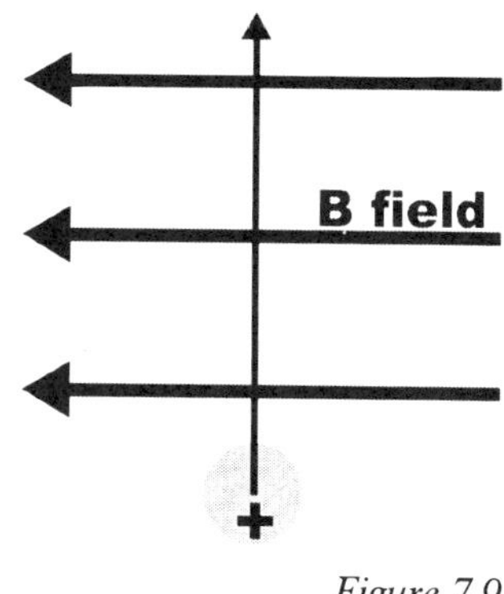

Figure 7.9

Step 1: Point the thumb of your right hand in the direction of the velocity (v) of a positive charge. If the picture shows a negative charge, point your thumb in the opposite direction of its motion.

The picture in Figure 7.9 shows a positive charge moving towards the top of the page, so point your thumb towards the top of the page.

Step 2: Point the index finger of your right hand in the direction of the magnetic field (B).

The picture shows the magnetic field is directed towards the left side of the page, so point your index finger in that direction.

Step 3: Point the middle finger of your right hand in a direction that is perpendicular to both the thumb and index finger of your right hand. The middle finger gives you the direction of the force (F).

Following this instruction, your middle finger should be pointing away from, or *out* of, the page. If this is not the case, go through the steps again and see if you can get it.

Answer: The magnetic force is directed out of the page.

Example 5

Look at Figure 7.10 and determine the force experienced by the current carrying wire.

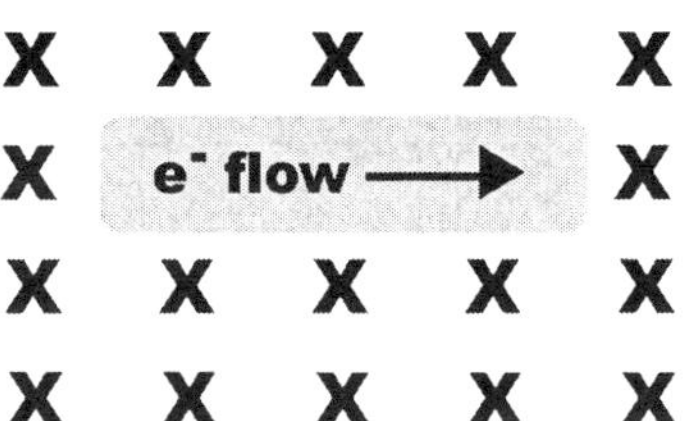

Figure 7.10

We handle wires in the same way that we handle individual charges.

Step 1: Point the thumb of your right hand in the direction of the velocity (v) of a positive charge. If the picture shows a negative charge, point your thumb in the opposite direction of its motion.

The picture shows the electron flow (negative charges) is to the right of the page, so reverse your thumb and point it to the left side of the page.

Step 2: Point the index finger of your right hand in the direction of the magnetic field (B).

The X symbols indicate that the magnetic field (B) is directed into the page, so point your index finger into the page.

Step 3: Point the middle finger of your right hand in a direction that is perpendicular to both the thumb and index finger of your right hand. The middle finger gives you the direction of the force (F).

Following this instruction, your middle finger should be pointing down, towards the bottom of the page.

Answer: The force is directed towards the bottom of the page.

Notice that our answers must be specific. Avoid using "up" or "down" as answers, because these terms aren't specific enough on their own, and your instructor or grader won't know if you mean "up, towards the top of the page" or "up, out of the page." Your answer must be specific and clear.

Notice that positively charged particles will experience a force that is in the opposite direction as the force experienced by a negatively charged particle with the same velocity. This allows physicists to use magnetic fields to separate particles. As a group of particles enter a magnetic field, positive charges will be deflected one way, negatively charged particles will be deflected in the opposite direction, and neutral particles will go straight through.

Lesson 7–3 Review

1. Imagine a magnetic field directed toward the left-hand side of the page, and a positive charge moving towards the bottom of the page. What would be the direction of the force experienced by this charge?
2. A 3.50 m wire segment with a steady current of 2.30 A through it experiences a force with a magnitude of 1.12 N when positioned perpendicular to a magnetic field. What is the magnitude of the magnetic field strength?
3. A charged particle with a velocity of 1.34×10^4 m/s experiences a force of 12.5 N as it travels through a magnetic field with a strength of 2.748×10^{-15} T at an angle of 45.0°. What is the charge on the particle?

Lesson 7–4: Electromagnetic Induction

As I mentioned in Lesson 7–2, Hans Christian Oersted showed that electricity could generate a magnetic field. Scientists then began to wonder if the reciprocal were also true. Could magnetic fields generate electricity? Through trial and error, two scientists eventually found that they could induce a current in a wire loop by moving a magnet back and forth through the center of it. Further experimentation showed that moving or turning a wire loop in a magnetic field will also induce a current. The two scientists were Michael Faraday (1791–1867), an Englishman, and an American named Joseph Henry (1797–1878). They share the credit for independently discovering electromagnetic induction, the process of generating electricity with a changing magnetic field. Very few discoveries, if any, have had a greater impact on our way of life.

Faraday, who is also responsible for the mental model of fields of force, concluded that when the number of magnetic field lines through the wire loop changes, a current is induced. Further, the magnitude of the induced current is proportional to the rate of this change. In other words, you can induce a stronger current in a wire loop by moving a bar magnet quickly through it. Because a stronger bar magnet would have more field lines per unit of area, you get more induced current with a stronger magnet. You could also increase the number of loops or turns in the wire coil. In summary, you can say that the current is induced by any change in **magnetic flux**.

Magnetic Flux

The product of the magnetic field and the area through which the magnetic field lines pass.

$$\Phi = BA\cos\theta$$

Where, Φ = the magnetic flux, B = the magnetic field, A = the area the field passes through, and θ = the angle between the field lines and the surface of the area they pass through.

Magnetic flux is measured in weber (Wb), where 1 Wb = 1 T · m^2.

Remember, Newton's second law tells us that the acceleration of an object is inversely proportional to its mass. Based on your understanding of Newton's second law, you will probably not be surprised to learn that it is often easier to turn a wire in a magnetic field than it is to move a bar magnet into and out of a wire coil. For this reason, many of today's generators employ this technique in order to induce current. Let's see how rotating a wire loop in an unchanging magnetic field can result in a change in magnetic flux.

Example 1

A square-shaped wire loop that is 3.0 cm long and 2.0 cm wide rests in a uniform 1.45 T magnetic field. The magnetic field is perpendicular to the surface of the area of the wire loop. Find the magnetic flux.

Convert: $L = 3.0\,\cancel{cm} \times \dfrac{1\,m}{100\,\cancel{cm}} = 0.030\,m$

$W = 2.0\,\cancel{cm} \times \dfrac{1\,m}{100\,\cancel{cm}} = 0.020\,m$

Calculate: $A = L \times W = (0.030\ m)(0.020\ m) = 6.0 \times 10^{-4}\ m^2$

Note, if the wire loop is perpendicular to the magnetic field, then the angle between the field and the surface of the area is zero.

Given: $B = 1.45\ T$ $\quad A = 6.0 \times 10^{-4}\ m^2$ $\quad \theta = 0^\circ$

Find: Φ

Solution:

$$\Phi = BA\cos\theta = (1.45\,T)(6.0 \times 10^{-4}\ m^2)(\cos 0^\circ)$$
$$= \mathbf{8.7 \times 10^{-4}\ Wb}$$

The value of cosine $0^\circ = 1$, so the magnetic flux through the wire loop is maximized when the field lines are perpendicular to the wire loop, meaning the angle between the field lines and the surface area is zero. What would happen if we took the same magnetic field and wire loop from Example 1 and rotated the wire loop so that the angle between the field lines and the surface of the area was 45°? Let's try that as our second example.

Example 2

A square-shaped wire loop that is 3.0 cm long and 2.0 cm wide rests in a uniform 1.45 T magnetic field. The angle between the field lines and the surface of the area of the wire loop is 45°. Find the magnetic flux.

Given: $B = 1.45\ T$ $\quad A = 6.0 \times 10^{-4}\ m^2$ $\quad \theta = 45^\circ$

Find: Φ

Solution:
$$\Phi = \text{BA}\cos\theta = (1.45\,\text{T})(6.0\times10^{-4}\ \text{m}^2)(\cos 45^\circ)$$
$$= \mathbf{6.2\times10^{-4}\ Wb}$$

As you can see in Example 2, changing the angle of the wire loop decreased the magnetic flux. This makes sense because the number of field lines actually going through the wire loop would have decreased. We can calculate the change in magnetic flux between Examples 1 and 2:

$$\Delta\Phi = \Phi_f - \Phi_i = (6.2\times10^{-4}\ \text{Wb}) - (8.7\times10^{-4}\ \text{Wb}) = -2.5\times10^{-4}\ \text{Wb}\,.$$

emf

I have avoided introducing **emf** until now, but we need to go over it in order to introduce Faraday's law of magnetic induction.

The letters *emf* actually stand for electromotive force. However, it turns out that the so-called "electromotive force" is not actually a force, so the initials emf are now more commonly used. Emf can be thought of as the voltage that is induced by magnetic induction. Emf is measured in volts (V) and is represented by the symbol ε.

Faraday's Law of Magnetic Induction

The instantaneous emf resulting from magnetic induction equals the rate of change of flux.

$$\varepsilon = -N\frac{\Delta\Phi}{\Delta t}$$

Where N = number of wire loops in a wire coil.

Example 3

Use Faraday's law of induction to determine the emf that would be generated in our previous examples, as we turned our wire loop 45° in 0.025 s.

Given: $\Delta\Phi = -2.5\times10^{-4}$ Wb N = 1 (a single wire loop)
$\Delta t = 0.025$ s

Find: ε

Solution: $\varepsilon = -N\dfrac{\Delta\Phi}{\Delta t} = \dfrac{-(1)(-2.5\times10^{-4}\ \text{Wb})}{0.025\ \text{s}} = \mathbf{0.010\ V}$

Notice that we divided a negative by a negative to get an answer with a positive sign. You might wonder about the significance of the sign in our answer. Before I explain it, let's go over the implications of what we just learned. You learned that charges in motion produce magnetic fields. You also learned that changing magnetic fields can generate emf, which will, in turn, generate electricity. This new electricity will generate a magnetic field, which as it springs into being, can produce an emf. You might be tempted to think that we could violate the law of conservation of energy in this way, but that is not the case. Lenz's law will stop us!

Lenz's Law

An induced emf gives rise to a current whose magnetic field opposes the change in magnetic flux that produced it.

What Lenz's law actually means is that we can't get something for nothing. If we move a wire loop in a magnetic field, we will generate current. However, the current will give rise to a magnetic field that will oppose the change that produced it. Lenz's law is the reason why we still need to put work into turning a wire in a magnetic field to generate electricity. Peoplc have come up with creative ways of turning generators, including making use of wind and falling water.

Lesson 7–4 Review

1. ______________ is the product of the magnetic field and the area through which the magnetic field lines pass.

2. A magnetic flux of 1.72 Wb results from a wire loop with an area of 2.00 m^2 orientated 35.0° from the field lines of a magnetic field going through it. Find the strength of the field.

3. How many wire loops would be required in a wire coil to generate an emf of 0.078 V when the change in magnetic flux through the coil is -7.5×10^{-4} Wb every 0.50 s?

Chapter 7 Examination

Part I—Matching

Match the following terms to the definitions that follow.

a. Hans Christian Oersted	d. weber	g. magnetic flux
b. Michael Faraday	e. volts	h. magnetic declination
c. Joseph Henry	f. tesla	i. magnetic domains

_____1. Emf is actually not a force, and it is not measured in newtons. It is actually measured in these units.

_____2. Magnetic flux is measured in these units.

_____3. Magnetic field strength is measured in these units.

_____4. This American scientist independently discovered magnetic induction.

_____5. This Danish scientist showed that an electric current generates a magnetic field.

_____6. The product of the magnetic field and the area through which the magnetic field lines pass.

Part II—Multiple Choice

For each of the following questions, select the best answer.

7. Which of the following is not a ferromagnetic element?
   a) iron b) nickel c) aluminum d) cobalt

8. A magnetic field will be produced by
   a) moving neutrons c) stationary electrons
   b) moving protons d) stationary neutrons

9. An electron (e^-) and an oxide ion (O^{2-}) are moving with the same velocity through a uniform magnetic field. Compared to the magnitude of the magnetic force experienced by the electron, the magnitude of the magnetic force experience by the oxide ion would be
   a) the same c) twice as great
   b) half as great d) four times as great

10. A charged particle moving through a magnetic field experiences a magnetic force. What is the angle between the magnetic field and the force on the particle?

    a) 0° b) 45° c) 90° d) 180°

11. An electron is moving through a magnetic field. If the velocity of the electron is doubled, the magnetic force experienced by the electron will

    a) stay the same c) triple
    b) double d) quadruple

12. Which law states, "An induced emf gives rise to a current whose magnetic field opposes the change in magnetic flux that produced it."

    a) Lenz's law c) Coulomb's law
    b) Faraday's law d) Newton's law

13. The instantaneous emf resulting from magnetic induction equals the rate of change of flux os

    a) Lenz's law c) Coulomb's law
    b) Faraday's law d) Newton's law

14. In the following formula, the "N" stand for $\varepsilon = -N\frac{\Delta\Phi}{\Delta t}$

    a) voltage c) number of moles
    b) emf d) number of loops

15. A wire loop is placed inside a magnetic field. The magnetic flux is maximized when the angle between the field lines and the surface of the area of the wire loop is

    a) 0° b) 45° c) 90° d) 180°

Part III—Calculations

Perform each of the following calculations.

16. An alpha particle (He^{2+}) with a velocity of 2.6×10^6 m/s crosses a uniform magnetic field at an angle of 37.0° to the field lines and experiences a magnetic force of 1.4×10^{-3} N. Find the strength of the magnetic field.

17. Find the magnetic force on a 3.50 m current carrying wire that is parallel to the field lines of a 2.30 T magnetic field, when the wire draws a current of 2.0 A.

18. Calculate the emf generated in a coil with 12 turns of wire and an area of 1.30 m^2 that was initially at rest in a uniform external magnetic field with a strength of 1.90 T, with the area of the loop orientated perpendicularly to the field lines, when it is pulled from the magnetic field in 0.30 seconds.

Answer Key

The actual answers will be shown in brackets, followed by an explanation. If you don't understand an explanation that is given in this section, you may want to go back and review the lesson that the question came from.

Lesson 7–1 Review

1. [magnetic domain]—In a magnetized substance, the magnetic domains are lined up in the same direction.
2. [ferromagnetic]—Recall, the magnetic properties of these elements are believed to result from the configurations of individual atoms.
3. [north]—If the magnetic south pole is located near the geographic North, it stands to reason that the magnetic north pole is located near the geographic South.

Lesson 7–2 Review

1. [south to north]—Outside of the solenoid, the lines come out of the south and into the north. Within the solenoid, they travel from the south pole to the north.
2. [counterclockwise]—Using the right-hand rule, if you hold the thumb of your right hand pointing out of the page, the other fingers of your right hand curl counterclockwise.
3. [into the page]—The thumb of your right hand would be pointing to the right, so the other fingers of your hand would go into the page below the wire, as you motioned to grab it.

Lesson 7–3 Review

1. [into the page]—Using the right-hand rule, we point the index finger of our right hand in the direction of the magnetic field, to the left of the page. We point our thumb in the direction of the velocity of the positive particle,

towards the bottom of the page. Our middle finger is at a right angle to both of the other fingers when it is directed into the page.

2. [0.139 T]—$B = \frac{F}{IL\sin\theta} = \frac{1.12\ \text{N}}{(2.30\ \text{A})(3.50\ \text{m})(\sin 90.0°)} = 0.139\ \text{T}$

3. [4.80 × 10[11] C]—$q = \frac{F}{Bv\sin\theta} = \frac{12.5\ \text{N}}{(2.748 \times 10^{-15}\ \text{T})(1.34 \times 10^{4}\ \text{m/s})(\sin 45.0°)}$
$= 4.80 \times 10^{11}\ \text{C}$

Lesson 7–4 Review

1. [magnetic flux]—The formula for calculating magnetic flux is $\Phi = BA\cos\theta$.

2. [1.05 T]— $B = \frac{\Phi}{A\cos\theta} = \frac{1.72\ \text{Wb}}{(2.00\ \text{m}^2)(\cos 35.0°)} = 1.05\ \text{T}$

3. [52 wire loops]—$N = \frac{-(\varepsilon)(\Delta t)}{\Delta\Phi} = \frac{-(0.078\ \text{V})(0.50\ \text{s})}{-7.5 \times 10^{-4}\ \text{Wb}} = 52$

Chapter 7 Examination

1. [e. volts]—emf is often simply represented with the symbol ε.
2. [d. weber]—Remember, 1 weber (Wb) = 1 T · m^2.
3. [f. tesla]—1 tesla (T) = 1 N/A · m
4. [c. Joseph Henry]—Henry also apparently invented the telegraph before Morse.
5. [a. Hans Christian Oersted]—The link between electricity and magnetism was established.
6. [g. magnetic flux]—Magnetic flux is represented with the symbol Φ.
7. [c. aluminum]—Aluminum foil is not strongly attracted to magnets.
8. [b. moving protons]—magnetic fields are produced by charges in motion. A neutron has no charge, so it won't produce a magnetic field.
9. [c. twice as great]—The formula, F = qvB, shows us that the magnetic force experienced by a moving charged particle in a magnetic field is proportional to the charge (q) on the object. The oxide ion has a charge that is twice as great as the charge on the electron.
10. [c. 90°]—the extended right-hand rule shows the direction of the force, in relation to the field.
11. [b. double]—The formula, F = qvB, indicates that the magnetic force experienced by a charged particle in motion through it is proportional to the velocity of the particle. If you double the velocity, you double the force.

12. [a. Lenz's law]—Lenz's law states that the law of conservation of energy is not violated by magnetic induction.
13. [b. Faraday's law]—More formally, this is called "Faraday's law of magnetic induction."
14. [d. number of loops]—The number of loops of wire in a coil.
15. [a. 0°]—Cosine 0° = 1, giving the maximum value for magnetic flux from the equation, $\Phi = BA\cos\theta$.
16. [2.8×10^9 T]—First, we can see that the alpha particle, which is a helium ion, has two excess protons. That gives the particle a charge (q) of ($q = n \times e = 2(+1.60 \times 10^{-19}\text{ C})) = +3.20 \times 10^{-19}$ C.
We then solve as shown.

$$B = \frac{F}{qv\sin\theta} = \frac{1.4 \times 10^{-3}\text{ N}}{(3.20 \times 10^{-19}\text{ C})(2.6 \times 10^{6}\text{ m/s})(\sin 37.0^{\circ})} = 2.8 \times 10^{9}\text{ T}$$

17. [0 N]—Be prepared for this type of question, which is designed to check if you are paying attention to the angle between the field lines and the wire. Sine 0º = 0, so the answer to our calculation will also be zero.

$$F = ILB\sin\theta = (2.0\text{ A})(3.50\text{ m})(2.30\text{ T})(\sin 0^{\circ}) = 0\text{ N}$$

18. [99 V]—That question is quite a mouthful, so you may have had trouble deciding where to start. The wire loop is being pulled out of the magnetic field, so the final magnetic flux will be zero. We need to find the initial magnetic flux to find the change in flux. Then we can use Faraday's law to determine the induced emf.

$$\Phi_i = BA\cos\theta = (1.90\text{ T})(1.30\text{ m}^2)(\cos 0^{\circ}) = 2.47\text{ Wb}$$

$$\Delta\Phi = \Phi_f - \Phi_i = (0\text{ Wb}) - (2.47\text{ Wb}) = -2.47\text{ Wb}$$

$$\varepsilon = -N\frac{\Delta\Phi}{\Delta t} = \frac{-12(2.47\text{ Wb})}{0.30\text{ s}} = 99\text{ V}$$

Waves and Light

One of the greatest debates in the history of physics concerned the nature of light. Was light made up of tiny particles, or did it travel in waves? The debate lasted for centuries, with scientists such as Sir Isaac Newton (1642–1727) and Albert Einstein (1879–1955) providing arguments for a particle theory, and Robert Hooke (1635–1703), Christian Huygens (1629–1695), and Thomas Young (1773–1829) favoring a wave theory. In this great debate, there was no clear winner. In the face of the evidence of numerous experiments, physicists have come to accept the wave-particle duality of light, meaning that light exhibits both wave-like and particle-like properties. In this chapter, we will take a closer look at the waves and also examine the wave-like properties of light.

Lesson 8–1: Types of Waves

To understand the wave-nature of light, we must discuss the properties and characteristics of waves first. We will start with a definition of waves, followed by a discussion of some of the types of waves.

A **wave** is defined as a disturbance that causes a transfer of energy over a distance, without a net transfer of mass. A type of wave with which you are probably familiar with is one that travels along the length of a taut rope when a person holding one end of it moves her hand up and down. Picture two people holding a length of rope between them. When one person holds his end steady and the other person moves her hand quickly up in a whip-like motion, a disturbance is created that will travel down the length of the rope until it reaches the other person. A single disturbance of this type is sometimes called a **pulse**. The person who was trying to hold his end steady will feel his arm jerk upwards, as the disturbance reaches him. In this way, you can imagine the energy that has been transferred along the length of the rope. The fact that the far end of the rope grows no thicker, over time, shows that there is no net transfer of mass.

If one person repeatedly moves her end of the rope up and down, she produces what is called a **continuous wave**, **periodic wave**, or **wave train**, as a series of disturbances will travel down the length of the rope. If the motion of the person is uniform, you will see an alternating pattern of high **crests** and low **troughs** traveling along the length of the rope.

These waveforms aren't limited to our example with ropes. If you keep your eyes open, you will see that these repeated patterns of disturbance occur in many materials, or media. To best understand a **material medium**, think about throwing a large rock into the water of a still pond. The rock strikes a certain spot, but the disturbance will spread out as ripples and affect a large area of the pond. A duck that was floating in the water several feet from where you threw the rock, would eventually start to bob up and down, as the wave disturbance passed by. The water, in this example, acts as a material medium, allowing the disturbance to spread through it.

One way to classify waves is to distinguish between **electromagnetic waves**, which are waves that can travel through a vacuum, and **mechanical waves**, which are waves that require a material medium through which to travel.

If you put an alarm clock or some other source of sound under the bell jar of a vacuum pump, you could still hear the alarm, as long as there was still air in the bell jar. The alarm would set up vibrations or waves in the air molecules surrounding it. The air would act as a material medium, transferring the disturbance to the glass jar, and vibrating its molecules. The molecules of the jar, acting as the new medium, would, in turn, transfer the disturbance to the molecules of air surrounding the jar. These air molecules would vibrate and pass the disturbance along until it reached your ears, disturbing the molecules of your eardrum.

What would happen if we turned the vacuum pump on and removed much of the air surrounding the clock, inside the bell jar? As we pumped the air molecules out of the jar, we would be removing the material medium that was passing the vibrations on to the molecules of the jar. We would notice the sound of the alarm diminishing and it would eventually disappear altogether. The reason is because sound waves are mechanical waves, and they require a medium through which to travel. Despite what you may see in movies or television, sound waves can't travel through outer space. If you were a spaceship pilot, and another spaceship flew quickly by your spaceship, you wouldn't hear anything.

Could you see another spaceship from the window of your spaceship? Can light travel through outer space? Of course it can! How else would the sun be able to light our way during the day, or the moon and stars by night? Light has no trouble traveling through a vacuum, so it doesn't fall under the definition of mechanical waves.

Early physicists attempting to defend the wave-theory of light "invented" an invisible material medium that surrounds the bodies of our universe, allowing them to argue that light was no different from other waves. They called this invisible medium that Earth floated in *ether*. The wave-theory of light was delivered a blow when an experiment designed to detect the ether failed to detect it. So a new category of waves that don't require a medium to travel was suggested. Light is an example of an electromagnetic wave.

There are many other forms of electromagnetic waves, which, together with visible light, make up the electromagnetic spectrum. Radio waves, for example, allow communication between spaceships and Earth. When an astronaut sends a message to Earth, the sound waves that he or she produces are converted to radio waves and beamed to the planet, where they are then converted back into sound waves. Other examples of electromagnetic waves are given in the following table.

Types of Electromagnetic Waves	
gamma rays	x-rays
ultraviolet light	visible light
infrared light	microwaves
radio waves	

Another way to categorize waves is to distinguish the direction in which they travel in relation to the vibrating particles of the medium in which they travel through. Think about the duck we disturbed when we threw a rock in the pond a few paragraphs ago. The waves spread out horizontally, in the form of concentric "ripples" along the surface of the water, but the duck bobs up and down perpendicularly to the disturbance that passes it.

In **transverse waves**, such as the ripples in the pond, the wave travels in a direction that is perpendicular to the displacement of the particles of the medium. Light and all of the electromagnetic waves are other examples of a transverse wave.

Longitudinal waves travel along the same axis as the disturbance of the particles. When you compress an area of a spring, and then release it, the disturbance travels down the length of the spring, along the same axis in which the molecules of the spring are being vibrated back and forth. Areas where the molecules of a medium become more tightly packed are called **compressions**. The areas where the molecules become less densely packed than normal are called **rarefactions**. Sound is an example of a longitudinal wave.

Notice that these categories of waves are not all mutually exclusive. Light waves are both transverse and electromagnetic because they travel in a direction that is perpendicular to the disturbance and they can travel without a material medium. Sound waves are both mechanical and longitudinal because they require a material medium through which to travel and they travel in a direction that is parallel to the disturbance of the medium.

Lesson 8–1 Review

1. A ______________ is a wave in which the particles of the medium vibrate back and forth along the same axis in which the wave travels.
2. A ______________ is a wave that requires a medium through which to travel.
3. A ______________ is an area in a material medium where the particles are less densely packed.

Lesson 8–2: Properties and Characteristics of Waves

Electromagnetic waves, such as light, differ from the waves that we are more familiar with in that they don't require a material medium through which to travel. In order to argue that light is made up of waves, people must have seen some properties or characteristics of waves that apply to light. Let's look at some of these features of light and waves in general.

Let's begin by talking about the anatomy of a wave. It may help to keep our example of the rope from Lesson 8–1 in mind as we talk about the "parts" of the wave. Keep in mind that waving a rope up and down produces a **transverse wave**, as the disturbance travels perpendicular to the displacement of the medium. The maximum displacement of the rope above its rest position, or the highest point on a wave, is called the *crest*. The maximum displacement of the rope below its rest position, or the lowest point on the wave, is called the *trough*. In a uniform wave, the trough is just as low as the crest is high, meaning that if you measure the distance from the rest position (equilibrium) of the wave to the trough it will be the same as the distance measured from the rest position to the crest. We call this distance of maximum displacement the **amplitude** of the wave.

We use the term *amplitude*, and also the terms that follow for both transverse and longitudinal waves. The compressions are analogous to the troughs and the rarefactions are analogous to the crests. So, for example, measuring the maximum displacement of the molecules in a rarefaction will still give you the amplitude of a longitudinal wave.

One complete waveform, meaning one crest and one trough in transverse wave, or one compression and one rarefaction in a longitudinal wave, makes up one **cycle**. If you measure the length of one complete cycle or waveform, it is called the **wavelength**. We will represent the wavelength with the Greek letter lambda (λ). It is important to remember that in a uniform wave train, the length of the crests and troughs are equal, so the length of each crest or trough is half of the wavelength. People will often measure from a certain point on one cycle to the same point on the next wave cycle to get the wavelength.

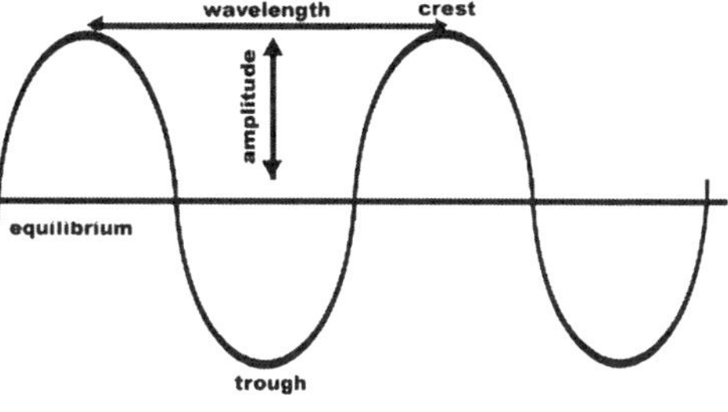

Figure 8.1

The amount of time it takes for one complete cycle (a crest and a trough) to pass by a given point is called the **period** (T) of the wave. So, for example, if our duck from our last lesson bobbed up and down through one complete crest and trough in 2.0 s, that would be the period of the wave. We discussed this concept in Chapter 4, when we talked about circular motion. The meaning of the term, period, is the same in this chapter. We are just using a more specific definition.

The number of complete waveforms, or cycles, that pass a given point in a second is called the **frequency** (f) of the wave. As you might recall from chapter four, the frequency of a wave is the inverse of its period, or

$$f = \frac{1}{T}.$$

So, if the period of the water wave that has been disturbing our poor duck is 2.0 s, then the frequency of the disturbance is

$$f = \frac{1}{T} = \frac{1}{2.0\,\text{s}} = 0.50\,\text{s}^{-1} = 0.50\,\text{Hz}.$$

Try and see past the numbers and units in these calculations to make sense of them. You will only be able to check to see if your answers make logical sense if you go beyond the numbers, and think of their actual meaning. If you take the example of the duck, realize that a period of 2.0 s means that one complete wave will go by every 2 seconds. We would then predict that half of a wave cycle should be able to go by in 1 second. In this way, our answer of 0.50 Hz, or 0.50 cycles/second, makes perfect sense.

Example 1

What is the period of a wave with a frequency of 5.0 Hz?

Given: f = 5.0 Hz

Find: T

Isolate:

Starting with: $f = \frac{1}{T}$

We multiply both sides by T: $f \times T = \frac{1}{T} \times T$

Then, divide both sides by f: $\frac{fT}{f} = \frac{1}{f}$

Giving us our working formula: $T = \frac{1}{f}$

Solution: $T = \frac{1}{f} = \frac{1}{5.0\ \text{Hz}} = \frac{1}{5.0\ \text{s}^{-1}} = \mathbf{0.20\ s}$

Does our answer to Example 1 make logical sense? If the wave has a frequency of 5.0 Hz, it means that every second, 5.0 complete cycles go by a given point. How long would it take for each cycle to go by that point? 1/5 of a second, making 0.20 s our period.

Speed of a Wave

The **speed of a wave** is obviously the amount of distance that a disturbance travels in a given period of time. As always, we could find the speed of a wave by using our old speed formula,

$$v = \frac{d}{\Delta t}.$$

However, we will often have access to the wavelength (λ) and period (T) of the wave, which are simply specific values of distance and time, so we often write our formula for the speed of a wave as

$$v = \frac{\lambda}{T}.$$

All electromagnetic waves travel at a speed of 3.00×10^8 m/s in a vacuum, and this constant is often represented with the letter C, making our formula

$$C = \frac{\lambda}{T}.$$

Example 2

In a vacuum, light travels at (C) 3.00×10^8 m/s. What would be the period of a ray of light with a wavelength of 590 nm (nanometers)?

Convert: $590\,\cancel{\text{nm}} \times \frac{1\text{ m}}{10^9\,\cancel{\text{nm}}} = 5.9 \times 10^{-7}\text{ m}$

Given: $C = 3.00 \times 10^8$ m/s $\lambda = 5.9 \times 10^{-7}$ m

Find: T

Isolate:

Start with: $C = \frac{\lambda}{T}$

We multiply both sides by T: $C \times T = \frac{\lambda}{\mathbf{T}} \times \mathbf{T}$

Divide both sides C: $\frac{\cancel{C}T}{\cancel{C}} = \frac{\lambda}{C}$

We get: $T = \frac{\lambda}{C}$

Solution: $T = \frac{\lambda}{C} = \frac{5.9 \times 10^{-7}\,\cancel{\text{m}}}{3.00 \times 10^8\,\cancel{\text{m}}/\text{s}} = \mathbf{2.0 \times 10^{-15}\ s}$

Our answer means that one complete cycle goes by every 0.000 000 000 000 00 20 seconds!

Going back to our example of waves traveling across the rope, what would happen if the person generating the waves moved her hand much faster? Would the speed at which a disturbance travels down the length of the rope be made to increase? The answer is no. The speed that a wave will travel through a medium is actually dictated by the medium, not by how rapidly the disturbances take place. If the person moved her hand faster, she would generate cycles faster by decreasing the wavelength of each waveform, and increasing the frequency of the waves. So, she would generate more, shorter waves, rather than make the waves travel faster. In other words, the frequency would go up, the wavelength would go down, but the speed would be the same.

To find the formula that shows the relationship between the frequency, wavelength, and speed of a wave, we can simply combine the two formulas that we have already used in this lesson:

Starting with $v = \frac{\lambda}{T}$ and $T = \frac{1}{f}$,

we substitute the value of T from the second formula into the first:

$$v = \frac{\lambda}{T} = \frac{\lambda}{\frac{1}{f}} = \lambda f.$$

This formula, $v = f\lambda$, is very useful, particularly when dealing with electromagnetic waves, where it is common practice to substitute the letter C (the speed of an electromagnetic wave in a vacuum) for v, making our formula, $C = f\lambda$.

Example 3

The operating frequency of a particular radio station is 93.3×10^6 Hz. What is the wavelength of the radio waves generated by this station?

Remember, radio waves are examples of electromagnetic waves, as they can travel through space. All electromagnetic waves travel at 3.00×10^8 m/s in space. Our atmosphere slows these waves so little that their speed in our atmosphere still rounds to 3.00×10^8 m/s, so we use this value as a constant for the speed of an electromagnetic wave in space or in our atmosphere.

$C = f\lambda$.

Given: $f = 93.3 \times 10^6$ Hz $C = 3.00 \times 10^8$ m/s

Find: λ

Solution: $\lambda = \frac{C}{f} = \frac{3.00 \times 10^8 \text{ m/s}}{93.3 \times 10^6 \text{ Hz}} = \mathbf{3.22\ m}$

Example 4

The speed of sound waves in air at a temperature of 20.0°C is approximately 340 m/s. Find the wavelength of a sound wave, under these conditions, with a frequency of 265 Hz.

Note, sound waves are *not* electromagnetic waves, so we don't use the constant (C).

Given: $v = 340$ m/s $f = 265$ Hz

Find: λ

Solution: $\lambda = \frac{v}{f} = \frac{340 \text{ m/s}}{265 \text{ Hz}} = \mathbf{1.3\ m}$

Interference

Is it possible to have multiple waves traveling through the same medium? Of course it is. Can you think of some examples from real life? When there is more than one source of sound in an area, there must be multiple sound waves traveling through the air. When more than one source is creating disturbances in the water, there will be multiple waves and ripples. What happens when waves occupy the same space?

The Superposition Principle of Waves

When two or more waves move through the same region of space, they maintain their individual integrity and will not change as a result of overlapping. While they occupy the same space, the individual disturbances will superimpose and produce a well-defined combined effect.

Constructive interference occurs when multiple waves, or parts of multiple waves, work together and generate a greater net disturbance than either of the individual disturbances. In this way, a number of smaller disturbances can add together to form a larger disturbance.

Destructive interference occurs when multiple waves, or parts of multiple waves, work against each other and generate a net disturbance that is smaller than the individual disturbances. In this way, a number of disturbances can add together and form a smaller disturbance. If two waves that have the same frequency and wavelength, but are 180° out of phase combine, the result can be **complete destructive interference**, where the waves cancel each other out in a specific area.

constructive interference

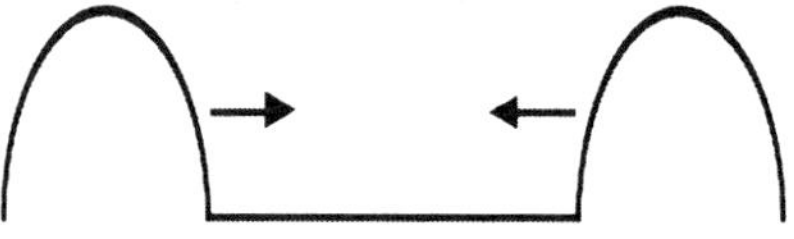

destructive interference

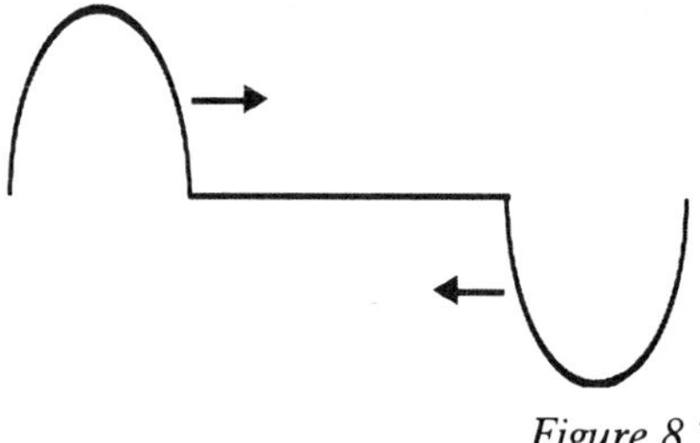

Figure 8.2

When Waves Strike the Boundary to a Medium

What happens when a wave hits a boundary to another medium? The wave can be **reflected**, meaning it will be turned back so that it will travel back through the medium in the opposite direction. Depending on the boundary to the medium that the wave strikes, the wave may be reversed or flipped over in such a way that it is 180° out of phase. The wave could also be **transmitted**, or allowed to pass into the next medium. Such a transmission may be associated with a **refraction**, which means that the wave can bend, or change directions, as it enters the new medium. We will go into more details about these characteristics in future lessons.

Another interesting property of waves, called **diffraction**, appears when a wave passes through an opening in a boundary to a medium. Waves will tend to spread out as they pass through such an opening. We can hear people speaking in another room, even if the door is closed, partially because the sound waves will spread out as they pass through the gap between the door and the floor.

Lesson 8–2 Review

1. _______________ is the addition of two or more waves resulting in a smaller net disturbance than the individual waves.
2. What is the wavelength of a microwave (in space) with a frequency of 2.40×10^9 Hz?
3. What is the period of an electromagnetic wave (in space) with a wavelength of 3.0×10^{-10} m?

Lesson 8–3: Sound Waves

As I mentioned earlier, sound waves are longitudinal mechanical waves. They require a medium to travel through, and the molecules of the medium will be displaced parallel to the direction in which the wave travels. It is also interesting to mention that sound waves radiate in three dimensions, allowing us to hear sources that are above us, below us, in front of us, behind us, etc.

Now that we have discussed many of the properties and characteristics of waves in general, it will help you to apply these to a specific type of wave that you are very familiar with, such as sound. In the last lesson, we gave an example of how sound waves exhibit diffraction, allowing sounds to spread out beyond barriers. Let's look at how some other properties of waves can be applied to sound. Making these real-life connections will help you understand and remember the many terms associated with waves.

Source of Sound	Approximate Intensity Level
Threshold of human hearing	0 dB
Whisper	20 dB
Normal speaking voice	60 dB
Vacuum cleaner	80 dB
Motorcycle	90 dB
Lawn mower	110 dB
Chain saw	120 dB
Threshold of pain	130 dB
Airplane	140 dB

Which property of waves would be associated with the loudness of a sound? How does a whisper differ from a shout? A loud noise represents a greater disturbance, in the same way that a large rock creates a larger disturbance than a small rock when thrown into a pool of water. Large rocks create bigger waves in the water, and loud noises create bigger waves in the air. In this case, when we say "bigger" waves, we are talking about the amplitude. The amplitude of a sound wave determines how loud we perceive it to be. The decibel (dB) scale is often associated with the amplitude or intensity of sound waves.

It is easy to imagine a higher frequency wave traveling through a rope or through water. High frequency would mean that many wave cycles would pass through a given point every second. How would our ears perceive a high frequency sound wave as opposed to a low frequency sound wave? We perceive higher frequency sound waves as high-pitched sounds, and lower frequency sound waves as low-pitched sounds.

Human ears can only detect sound waves in a specific range—from around 20 to 20,000 hertz. As you may know, dogs can hear high-pitched sounds that humans can't. A dog whistle that produces sound waves with a frequency of 35,000 hertz is quite inaudible to the human ear, yet still well within the hearing range of dogs. Bats are capable of producing and hearing sounds that are too high-pitched for even a dog to hear them, giving them the **echolocation**, or "radar," for which they are famous.

How would you explain an echo in terms of waves? An echo is a result of the characteristic of waves called reflection. When a wave hits a certain barrier, such as a wall, it can be reflected or turned back. A bat can produce a high-pitched sound and if it hears that it is reflected back very quickly, it knows there is something in front of it.

Much like a bat, your ears can give you information about nearby objects that you may not even pay attention to. You probably use information about the amplitude of sound waves to determine if an object is coming towards you. For example, an approaching airplane or truck seems to grow louder as it grows closer. What you might not realize is that you can often tell if the source of a sound is getting closer to you by listening for a change in the pitch of the sound.

The Doppler Effect

The apparent change in frequency or wavelength of a wave that is perceived by an observer moving relative to the source of the waves.

The Doppler effect is responsible for the change in the pitch of a sound as it moves relative to you. When the source of a sound is approaching you, the sound waves that it produces are reaching you faster, resulting in a higher perceived frequency and a higher perceived pitch. When a sound source is moving away from you, the sound waves reach you at a slower rate, resulting in a lower pitched sound. The next time you hear a siren in the street, try to listen for changes in pitch, as well as changes in amplitude.

Lesson 8–3 Review

1. ______________ is the apparent change in frequency or wavelength of a wave that is perceived by an observer moving relative to the source of the waves.
2. The ______________ of a sound wave is responsible for its perceived loudness or intensity.
3. The higher the ______________ of a sound wave, the higher the perceived pitch.

Lesson 8–4: Light

Now that you know some of the properties of waves, you can think about how these properties are exhibited by light. This is just what earlier scientists did during the particle-wave debate. If you want to argue that light travels in waves, you need to show that light exhibits wave-like properties.

Anyone who has looked in a mirror or the surface of a still body of water must accept the fact that light exhibits the property called **reflection**. As light strikes a surface, some portion of it will be turned back. Unfortunately, this doesn't really prove that light is made up of waves, because particles exhibit reflection as well. Understanding this is what allows billiards players to make bank shots.

Light can also be **refracted** when it passes obliquely (at an angle less than 90°) into a new medium. However, it is easy to imagine how the path of a *particle* might get refracted as well. For example, if a golf cart that was driving on a paved pathway passed onto the wet grass of a golf course at a 45° angle, the tires on one side of the cart might slip first, causing the cart to turn slightly. So, the fact that light refracts was not considered conclusive evidence to support a wave-theory of light.

For a time, the fact that light did not seem to exhibit **diffraction** was used as evidence to argue against the wave-theory of light. If your teacher were to duck behind her desk and keep talking, you would still hear her, because sound waves diffract, or spread out, beyond the barrier of the desk. You would not be able to see her, however, so light doesn't seem to diffract in this case. However, it was eventually shown that light *does* diffract if you use an opening or barrier that is sufficiently small. If you block a light source with a very small object, the object can be shown to cast a shadow even smaller than itself, showing that light spreads out beyond it.

It was also once thought that light didn't show the characteristic of waves called **interference**. However, an English scientist named Thomas Young performed a conclusive experiment in 1801 called the "double-slit" experiment, which showed that light can be made to exhibit both diffraction and interference. Finding no other way to interpret the results of the double-slit experiment, the skeptical scientists were forced to accept the wave-nature of light.

The story doesn't end there, because in 1905, a 26-year-old patent clerk published a paper arguing that the photoelectric effect, which occurs when light shines on a piece of metal and causes electrons to be ejected, can only be fully understood if you think of light as being made up of particles. Finding no better interpretation for the photoelectric effect, scientists eventually accepted the wave-particle duality of nature, meaning that light has both wave-like and particle-like properties. The young patent clerk received the Nobel Prize for Physics. His name? Albert Einstein.

No review questions this time. Just move on to the next lesson.

Lesson 8–5: Reflection and Mirrors

When light strikes an object, some portion of it is absorbed and some is reflected off the surface. Some objects absorb most of the light striking their surface, and they appear dark-colored to our eyes. When an object appears to be light-colored or white, much of the light is being reflected off the surface.

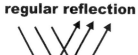

What is the difference between an object that appears to be a brilliant white, and an object with a reflective surface, such as a mirror? The surface of a mirror is so smooth that it produces a **regular reflection**. A regular reflection occurs when light goes towards the surface in parallel, **incident rays**, and bounces off the surface in parallel, ***reflected* rays**. A brilliant white object can reflect as much or nearly as much light as a mirror, but the surface is not as smooth, so the reflected rays will not be parallel. Such a reflection is called an **irregular**, or **diffuse**, **reflection**.

regular reflection

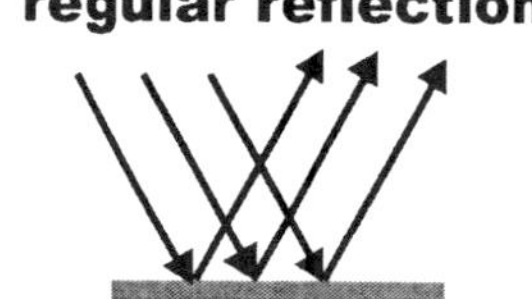

diffuse reflection

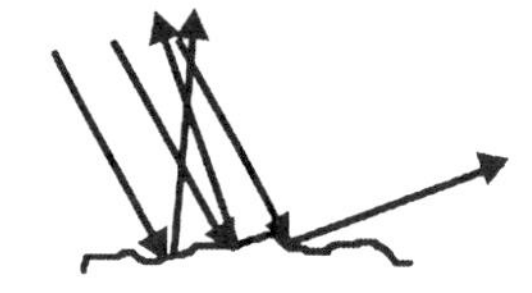

Figure 8.3

It is important to remember that it is the nature of the surface, not the nature of the light striking the surface, that determines the type of reflection that occurs. All light rays follow what is called the **law of reflection**. The law of reflection states that the **angle of incidence** will always be equal to the angle of reflection. The angle of incidence is formed between the incident (incoming) ray of light and the normal (perpendicular) to the medium boundary. The angle of reflection is the angle between the reflected (outgoing) ray and the normal.

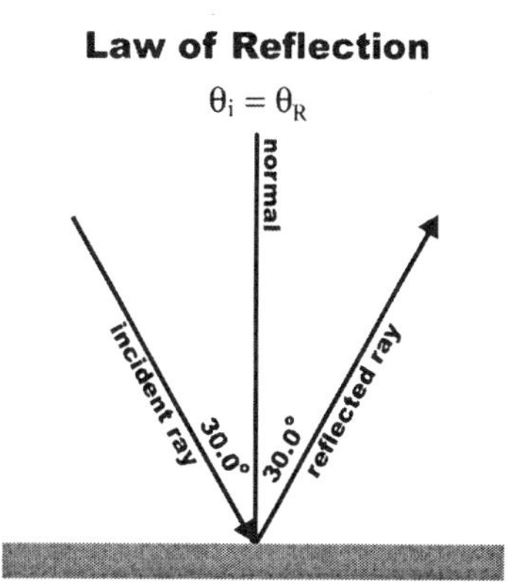

Figure 8.4

A diffuse reflection still follows the law of reflection; however, because the surface is rough, the normal to the surface will be different at each point, resulting in the reflected rays being scattered.

The light striking curved mirrors must also follow the law of reflection, but just as with our rough surface, the normal to the mirror will be different at every spot. The resultant reflections off such mirrors are distorted in ways that depend on a number of factors, such as the focal length of the mirror and the distance of the object from the mirror.

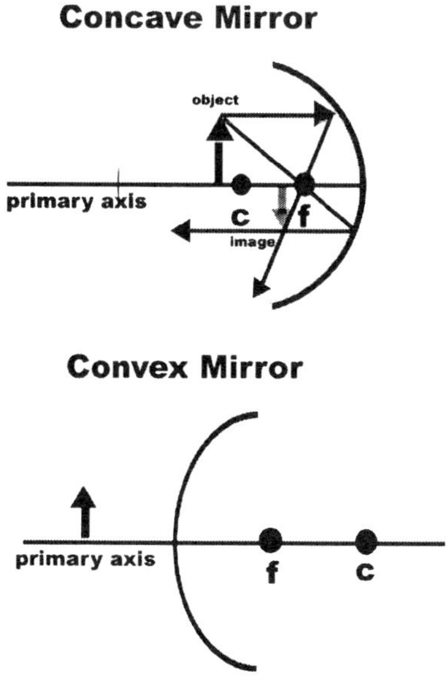

Concave Mirror

object
primary axis
c
f
image

Convex Mirror

primary axis
f
c

Figure 8.5

Two types of curved mirrors that you are likely to study this year are the **concave** (converging) and **convex** (diverging) mirrors. Both of these mirrors are curved, but the shiny surface is on the inner curve of a concave mirror and the outer curve of a convex mirror. Looking at the inside of a shiny metal spoon is similar to looking into a concave mirror. Looking at the back of a shiny metal spoon is similar to looking at a convex mirror. Concave, or "converging," mirrors are used in searchlights because they help focus the rays of light into smaller and, therefore, brighter areas. Convex, or "diverging," mirrors are used for side-view mirrors and in convenience stores because they reflect light from a wide area, resulting in a larger field of view.

If you imagine the curved mirror as being a slice taken from a large sphere or ball, the **center of curvature** (C) would represent the center of that sphere. This places the center of curvature in front of a concave mirror, and behind a convex one. The center of curvature is located at a distance that is twice the **focal length** of a particular mirror.

Halfway between the center of curvature and the actual **vertex**, or center of the mirror, lies the **focal point** (f). Again, this places the focal point for a concave mirror in front, and for a convex mirror, behind. A line drawn through the center of curvature, focal point, and the vertex of the mirror is called the **primary axis**. For a concave mirror, the focal point represents the point that all light that comes in parallel to the primary axis gets reflected through. For a convex mirror, the focal point is where the reflected light seems to radiate from.

Curved mirrors are capable of producing a variety of images, depending on how far from the mirror the reflected object is placed. Some images are upright, meaning right-side-up. Other images are inverted, or upside-down. Some images are enlarged or bigger, others are reduced or smaller, still others appear to be the same size as the original object. Some images, which are formed by converging rays of light, are called **real images**. Other images, formed by diverging rays of light, are called **virtual images**.

Images Formed by Concave Mirrors

1. When you place an object beyond the center of curvature, the image that is produced will be real, inverted, and smaller than the object.

2. If you place the object at the center of curvature, the image will still be real and inverted, but it will appear to be the same size as the object.

3. If you place the object at a distance that is between the center of curvature and the focal point, the object will be real, inverted, and larger than the object.

4. If you place the object at the focal point, the reflection will appear as a blur. No image is formed. You can use this effect to locate the focal length of a mirror.
5. If you place the object between the focal point and the surface of the mirror, the image will be virtual, upright, and larger than the object.

Images Formed by Convex Mirrors

All images formed by convex mirrors are virtual, upright, and smaller than the object.

Sometimes, drawing a ray diagram, such as the one shown in Figure 8.5, is the most convenient way of determining the type of image that will be formed in a specific situation. However, sometimes you will be asked to find out more specific information, such as the actual distance that the image will be formed from the mirror, the height of the image, or the focal length of the mirror. In such situations, the following equations come in handy.

Equations for Curved Mirrors

Height Equation →

$$\frac{\text{height of the image}}{\text{height of the object}} = -\frac{\text{distance of the image}}{\text{distance of the object}}$$

or

$$\frac{h_i}{h_o} = -\frac{d_i}{d_o}$$

Distance Equation →

$$\frac{1}{\text{focal length}} = \frac{1}{\text{distance of object}} + \frac{1}{\text{distance of image}}$$

or

$$\frac{1}{f} = \frac{1}{d_o} + \frac{1}{d_i}$$

The signs involved with these equations can be somewhat tricky, so you will want to pay attention to the following sign conventions.

Concave (Converging) Mirror

- The focal length (f) is always positive.
- The distance of the object (d_o) is always positive.
- The distance of the image (d_i) is positive for a real image, formed in front of the mirror, and negative for a virtual image, formed behind the mirror.
- The height of the image (h_i) will be positive if the image is upright, and negative if it is inverted.

Convex (Diverging) Mirror

- The focal length (f) is always negative.
- The distance of the object (d_o) is always positive.
- The distance of the image (d_i) is always negative, because only a virtual image can be formed by this type of mirror.
- The height of the image (h_i) will always be positive, because all images formed by this mirror will be upright.

Example 1

A candle is placed 14.0 cm in front of a concave mirror with a focal length of 10.0 cm. Find the distance from the mirror that the image will be formed.

The first thing that we do is go back and check our sign conventions for concave mirrors, to see if we need to make any of our "givens" negative. The check reveals that for concave mirrors, both of our givens should be positive.

Given: $d_o = 14.0$ cm $\qquad f = 10.0$ cm

Find: d_i

Isolate:

Starting with our original equation: $\frac{1}{f} = \frac{1}{d_o} + \frac{1}{d_i}$

We subtract $\frac{1}{d_o}$ from both sides, giving us: $\frac{1}{d_i} = \frac{1}{f} - \frac{1}{d_o}$

Finding a common denominator for the right side gives us: $\frac{1}{d_i} = \frac{d_o - f}{d_o f}$

Taking the inverse of both sides gives us our working formula:

$$d_i = \frac{d_o f}{d_o - f}$$

Solution: $d_i = \frac{d_o f}{d_o - f} = \frac{(14.0\text{ cm})(10.0\text{ cm})}{14.0\text{ cm} - 10.0\text{ cm}} = \mathbf{35.0\text{ cm}}$

Checking back to our sign conventions for concave mirrors, we can see that our positive value for the image distance describes a real image, formed in front of the mirror. It will be inverted, because all real images are inverted, and it will be enlarged as well, because it is formed beyond the object.

Example 2

How far away from a convex mirror should an object with a height of 6.0 cm be placed in order to form a virtual image with a height of 4.0 cm at a distance of 2.0 cm behind the mirror?

The question didn't need to tell us that the image was virtual, because convex mirrors only form virtual images. Checking our sign conventions for convex mirrors, we are reminded that the distance of the image (d_i) formed by these mirrors will always be negative, because they are formed behind the mirror.

Given: $h_o = 6.0$ cm $h_i = 4.0$ cm $d_i = -2.0$ cm

Find: d_o

Isolate:

Starting with: $\frac{h_i}{h_o} = -\frac{d_i}{d_o}$

We multiply both sides by d_o: $d_o \times \frac{h_i}{h_o} = \frac{-d_i}{d_o} \times d_o$

Multiply both sides by h_o: $h_o \times \frac{d_o h_i}{h_o} = -d_i \times h_o$

Divide both sides by h_i: $\frac{d_o h_i}{h_i} = \frac{-d_i h_o}{h_i}$

Leaving us with: $d_o = \frac{-d_i h_o}{h_i}$

Solution: $d_o = \frac{-d_i h_o}{h_i} = \frac{-(-2.0\text{ cm})(6.0\text{ cm})}{4.0\text{ cm}} = \mathbf{3.0\text{ cm}}$

The positive value of our answer indicates that the object is placed in front of the mirror.

Example 3

A candle with a height of 3.5 cm is placed at a distance of 12.0 cm in front of a concave mirror that has a focal length of 10.0 cm. Find the distance at which the image is formed from the mirror and the height of the image.

Checking my sign conventions for a concave mirror, I don't see anything that needs to be given a negative value.

Given: $h_o = 3.5$ cm $d_o = 12.0$ cm $f = 10.0$ cm

Find: d_i and h_i

Solution: $d_i = \frac{d_o f}{d_o - f} = \frac{(12.0\text{ cm})(10.0\text{ cm})}{(12.0\text{ cm} - 10.0\text{ cm})} = \mathbf{60.0\text{ cm}}$

$h_i = \frac{-d_i h_o}{d_o} = \frac{-(60.0\text{ cm})(3.5\text{ cm})}{12.0\text{ cm}} = \mathbf{-17.5\text{ cm}}$

The negative sign for our second answer indicates that the image is inverted.

Lesson 8–5 Review

1. A ______________ is an image that is formed by diverging rays of light.
2. A lightbulb with a height of 8.5 cm is placed at a distance of 8.0 cm in front of a concave mirror that has a focal length of 14.0 cm. Find the distance that the image is formed from the mirror and the height of the image.
3. A candle with a height of 5.0 cm is placed at a distance of 22.0 cm in front of a convex mirror that has a focal length of 8.0 cm. Find the distance that the image is formed from the mirror and the height of the image.

Lesson 8–6: Refraction and Lenses

Refraction makes objects appear to be bent when placed in water. If you ever used a net on a pole to clean leaves out of the bottom of a pool, or a net to remove a fish from a fish tank, you probably noticed that these items appear to bend right at the surface of the water. If you haven't had such an experience, you have probably noticed that people's legs appear very squat when they are standing in a body of water, when viewed from outside the water. Refraction involves the bending of waves as they enter a new medium.

The key to understanding how and why refraction occurs begins with a discussion of material media. Light (and all forms of electromagnetic waves) has a maximum speed of 3.00×10^8 m/s in a vacuum. When light enters a new medium, its speed changes based on the **index of refraction** (n) of that particular medium. The index of refraction (n) is simply the ratio of the speed of light in a vacuum (C) to the speed of light in that particular medium (v).

Calculating the Index of Refraction for a Medium

$$\text{index of refraction of a medium (n)} = \frac{\text{speed of light in a vacuum (C)}}{\text{speed of light in the medium } (v)}$$

or

$$n = \frac{C}{v}$$

Light from the sun slows down when it hits our atmosphere, but it slows down by such a small amount that the speed of light in our atmosphere still rounds to 3.00×10^8 m/s. So, the index of refraction for air (n_{air}) is equal to

$$n_{air} = \frac{C}{v} = \frac{3.00 \times 10^8 \cancel{m/s}}{3.00 \times 10^8 \cancel{m/s}} = 1.00 .$$

Notice that all of the units are cancelled in our calculation, so the index of refraction has no units.

Example 1

Light has a speed of 1.80×10^8 m/s in a particular piece of glass. Calculate the index of refraction for this glass.

Given: $C = 3.00 \times 10^8$ m/s $\qquad v = 1.80 \times 10^8$ m/s

Find: n

Solution: $n = \frac{C}{v} = \frac{3.00 \times 10^8 \cancel{m/s}}{1.80 \times 10^8 \cancel{m/s}} = \mathbf{1.67}$

Example 2

The index of refraction for a sample of water is 1.33. How fast does light travel through this water?

Given: n = 1.33 $\quad C = 3.00 \times 10^8$ m/s

Find: v

Solution: $v = \frac{C}{n} = \frac{3.00 \times 10^8 \text{ m/s}}{1.33} = \mathbf{2.26 \times 10^8 \text{ m/s}}$

Recall our example of a golf cart driving at a 45° angle into slippery grass, and turning slightly as the wheels on one side of the cart slip. When light passes at an angle into a new medium, it will bend because one side of the wave changes speed before the other side. The direction the light will bend towards when it enters at an angle into a new medium is given by the **law of refraction**.

The Law of Refraction

When light passes at an angle from a medium with a lower index of refraction to one with a higher index of refraction, it will bend towards the normal to the boundary.

When light passes at an angle from a medium with a higher index of refraction to one with a lower index of refraction, it will bend away from the normal to the boundary.

The extent to which a wave of light bends when it passes obliquely into a new medium, is given by Snell's law.

Snell's Law

$$n_1 \sin \theta_1 = n_2 \sin \theta_2$$

Where θ_1 is the angle of incidence, and θ_2 is the angle of refraction.

Example 3

A ray of light passes through air and strikes a piece of glass at an angle of 45.0° from the normal. If the index of refraction for the glass is 1.60, what will be the angle of refraction?

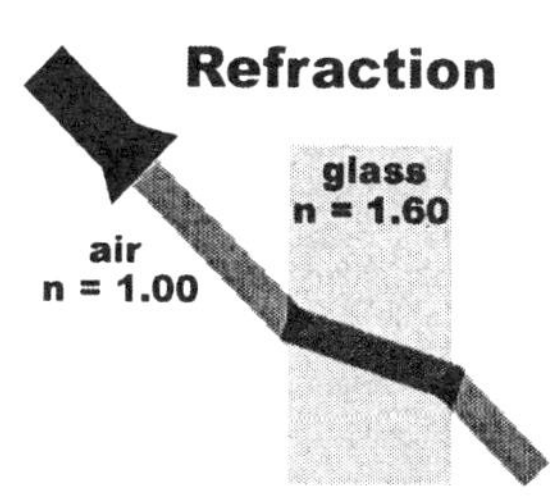

Given: $n_1 = 1.00$ $\qquad \theta_1 = 45.0°$

$n_2 = 1.60$

Find: θ_2

Isolation:

Starting with: $n_1 \sin \theta_1 = n_2 \sin \theta_2$

Divide both sides by n_2:

$$\frac{n_1 \sin \theta_1}{n_2} = \frac{\cancel{n}_2 \sin \theta_2}{\cancel{n}_2}$$

Rearranging, we get: $\sin \theta_2 = \dfrac{n_1 \sin \theta_1}{n_2}$

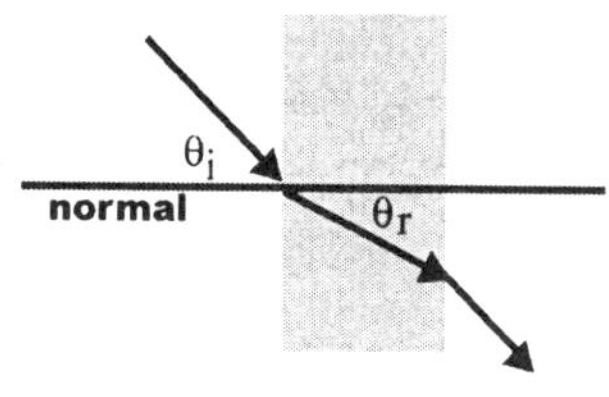

Figure 8.6

Solving for θ_2, we get: $\theta_2 = \sin^{-1}\left(\frac{n_1 \sin\theta_1}{n_2}\right)$

Solution: $\theta_2 = \sin^{-1}\left(\frac{n_1 \sin\theta_1}{n_2}\right) = \sin^{-1}\left(\frac{(1.00)(\sin 45.0°)}{1.60}\right) = \mathbf{26.2°}$

Example 4

Light passes through air ($n_1 = 1.00$) and strikes the surface of a piece of glass at an angle of 58.0° from the normal. The angle of refraction is 33.0° measured from the normal. Calculate the index of refraction for the glass.

Given: $n_1 = 1.00$ $\theta_1 = 58.0°$ $\theta_2 = 33.0°$

Find: n_2

Isolate:

Starting with: $n_1 \sin\theta_1 = n_2 \sin\theta_2$

Divide both sides by $\sin\theta_2$: $\frac{n_1 \sin\theta_1}{\sin\theta_2} = \frac{n_2 \cancel{\sin\theta_2}}{\cancel{\sin\theta_2}}$

Rearranging, we get: $n_2 = \frac{n_1 \sin\theta_1}{\sin\theta_2}$

Solution: $n_2 = \frac{n_1 \sin\theta_1}{\sin\theta_2} = \frac{(1.00)(\sin 58.0°)}{(\sin 33.0°)} = \mathbf{1.56}$

Critical Angle of Refraction

Did you ever look through a fish tank containing water and notice that one or two of the other sides of the tank look like mirrors? You can look straight through a fish tank and see that the back wall is actually clear glass, and yet when you look through from the side, you can no longer see through the back wall. In such a situation, you are observing a phenomenon known as **total internal reflection**.

One of the implications of the law of refraction is that when light passes from a medium with a greater index of refraction to a medium with a lower index of refraction, the angle of refraction will be bigger than the angle of incidence. That means that if the angle of incidence is large enough, the angle of refraction would be made to exceed a 90° angle! Such a ray would not refract into the next medium at all, but would reflect back into the original medium. The effect, total internal reflection, occurs when the **critical angle of incidence** has been exceeded.

The formula for calculating the critical angle of incidence is:

$$\sin\theta_c = \frac{n_2}{n_1}.$$

Because the index of refraction of air is 1.00, when the second medium involved in a problem is air, this equation may simply be written as:

$$\sin\theta_c = \frac{1}{n_1}.$$

Example 5

The index of refraction for water is 1.33. Above the water is air (n = 1.00). What is the critical angle of incidence for light starting inside the water and passing into the air above it?

Given: $n_1 = 1.33 \quad n_2 = 1.00$

Find: θ_c

Isolate:

Starting with: $\sin\theta_c = \dfrac{n_2}{n_1}$

We solve for θ_c: $\theta_c = \sin^{-1}\left(\dfrac{n_2}{n_1}\right)$

Solution: $\theta_c = \sin^{-1}\left(\dfrac{n_2}{n_1}\right) = \sin^{-1}\left(\dfrac{1.00}{1.33}\right) = \mathbf{48.8^\circ}$

Example 6

A diamond with an index of refraction of 2.42 is surrounded by air ($n = 1.00$). If a ray of light originating inside the diamond strikes the surface of a facet at an angle of 27.0° from the normal, will the ray be refracted, or will it experience total internal reflection?

The key to this problem is to figure out if the angle of incidence (27.0°) exceeds the critical angle for total internal reflection. If it is greater than the critical angle, the ray is reflected. If it is less than the critical angle, it will be refracted according to Snell's law.

Given: $n_1 = 2.42 \quad n_2 = 1.00 \qquad \theta_i = 27.0°$

Find: θ_c

Solution: $\theta_c = \sin^{-1}\left(\frac{n_2}{n_1}\right) = \sin^{-1}\left(\frac{1.00}{2.42}\right) = \mathbf{24.4°}$

Because the angle of incidence exceeds the critical angle, the light will experience total internal reflection.

The property of *refraction* is what allows lenses to be used in eyeglasses to correct vision problems. It is also what allows lenses to be used to magnify images in telescopes, binoculars, and microscopes. Lenses can be ground into many different types of shapes, but we will limit our discussion here to two major types of lenses, *convex* and *concave*.

Convex (converging) Lens

Concave (diverging) Lens

Figure 8.7

Convex lenses are called converging lenses, because incident rays that are parallel to the primary axis will be refracted through the focal point (f). Incident rays that pass through the secondary focus (f′) will be refracted in such a way that they end up parallel to the primary axis. The result is that rays of light that pass through the lens are made to converge at a point. This type of lens can be used as a magnifying glass and to start fires.

Concave lenses are called diverging lenses because light that passes through such a lens will refract away from the focal point. This type of lens would not help you start a fire, but it can be used to correct some types of vision problems.

The following ray diagram (Figure 8.8) illustrates why convex lenses are called converging lenses. Notice that the diagram includes a notable simplification. In reality, the ray of light will refract twice. Once when it enters the lens, and once when it leaves the lens. Most ray diagrams are simplified to show only one refraction, but you should always remember that this represents the net effect of two refractions.

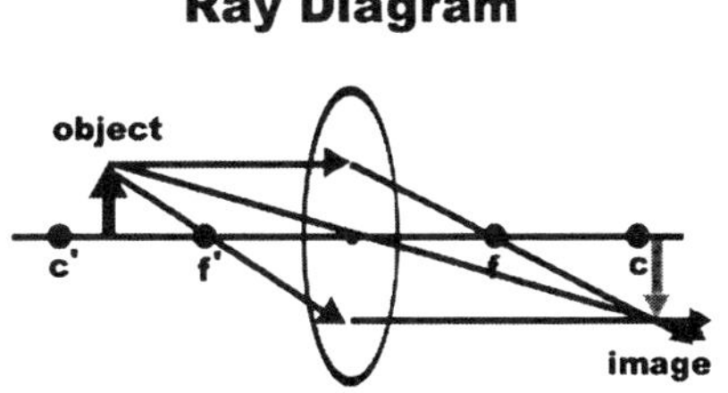

Figure 8.8

You can see from the ray diagram that we use many of the same concepts for lenses that we use for mirrors. We still talk about the primary axis, vertex, center of curvature, focal length, and focal point. In fact, as you will see shortly, we even use some of the same equations for lenses that we used for mirrors in our previous lesson.

Equations for Thin Lenses

Height Equation → $\frac{\text{height of the image}}{\text{height of the object}} = \frac{\text{distance of the image}}{\text{distance of the object}}$

or

$$\frac{h_i}{h_o} = -\frac{d_i}{d_o}$$

Distance Equation →

$$\frac{1}{\text{focal length}} = \frac{1}{\text{distance of object}} + \frac{1}{\text{distance of image}}$$

or

$$\frac{1}{f} = \frac{1}{d_o} + \frac{1}{d_i}$$

Although the equations are identical, we must pay careful attention to another set of sign conventions when dealing with lenses.

Concave (Diverging) Lens

- The focal length (f) is always negative.
- The distance of the object (d_o) is always positive.
- The distance of the image (d_i) is always negative, because only a virtual image can be formed by this type of lens.
- The height of the image (h_i) will always be positive, because all images formed by this lens will be upright.

Convex (Converging) Lens

- The focal length (f) is always positive.
- The distance of the object (d_o) is positive if the object is placed in front of the lens, and negative if the object is placed behind the lens.
- The distance of the image (d_i) is positive for a real image, which is formed on the opposite side of the lens as the object, and negative for a virtual image, which is formed on the same side of the lens as the object.
- The height of the image (h_i) will be positive if the image is upright, and negative if it is inverted.

Example 7

A bulb with a height of 6.00 cm is placed 24.0 cm in front of a convex lens with a focal length of 16.0 cm. Calculate the height of the image formed and the distance the image will be formed from the lens.

Checking the sign conventions for convex lenses reveals that we don't need to list any of our given values with negative signs.

Given: $h_o = 6.00$ cm $d_o = 24.0$ cm $f = 16.0$ cm

Find: d_i and h_i

Solution: $d_i = \dfrac{d_o f}{d_o - f} = \dfrac{(24.0\text{ cm})(16.0\text{ cm})}{(24.0\text{ cm} - 16.0\text{ cm})} = \mathbf{48.0\text{ cm}}$

The fact that the value for the distance of the image is positive means that it forms behind the lens. It is a real image.

$$h_i = \frac{-d_i h_o}{d_o} = \frac{-(48.0\text{ cm})(6.00\text{ cm})}{24.0\text{ cm}} = \mathbf{-12.0\text{ cm}}$$

The fact that the height of the image has a negative value means that the image is inverted. It is also larger than the object.

Example 8

A candle with a height of 4.00 cm is placed 14.0 cm in front of a concave lens with a focal length of 18.0 cm. Calculate the height of the image formed and the distance the image will be formed from the lens.

Checking the sign conventions for concave lenses reveals that our focal length should be given a negative sign.

Given: $h_o = 6.00$ cm $\quad$ $d_o = 14.0$ cm $\quad$ $f = -18.0$ cm

Find: d_i and h_i

Solution: $d_i = \dfrac{d_o f}{d_o - f} = \dfrac{(14.0\text{ cm})(-18.0\text{ cm})}{(14.0\text{ cm}) - (-18.0\text{ cm})} = \mathbf{-7.88\text{ cm}}$

As is always the case with concave lenses, the distance of the image is negative, placing the image on the same side of the lens as the object.

$$h_i = \frac{-d_i h_o}{d_o} = \frac{-(-7.88\text{ cm})(6.00\text{ cm})}{14.0\text{ cm}} = \mathbf{3.38\text{ cm}}$$

The height of the image is positive, because concave lenses always form upright images. The image is also reduced.

Lesson 8–6 Review

1. A diamond has an index of refraction of 2.42. Calculate the speed of light in the diamond.

2. A ray of light passes through air and strikes a smooth body of water at an angle of 22.0° from the normal. If the index of refraction for the water is 1.33, what will be the angle of refraction?

3. An object with a height of 5.00 cm is placed 22.0 cm in front of a concave lens with a focal length of 11.0 cm. Calculate the height of the image formed and the distance the image will be formed from the lens.

Chapter 8 Examination

Part I—Matching

Match the following terms to the definitions that follow.

a. crest	e. mechanical wave	i. frequency
b. trough	f. electromagnetic wave	j. compression
c. wavelength	g. transverse wave	k. convex
d. amplitude	h. longitudinal wave	l. concave

_____1. A transverse wave that is capable of traveling through a vacuum.

_____2. An area in a material medium where the particles are more densely packed.

_____3. The distance measured from a point on one cycle to the same point on the next cycle.

_____4. The number of vibrations or cycles of motion in a second, usually measured in hertz.

_____5. The maximum displacement of the particles of a medium from their rest position.

_____6. A wave in which the particles of the medium vibrate back and forth along the same axis in which the wave travels.

_____7. The type of curved mirror with the shiny surface on the outside of the curve.

_____8. A wave that requires a medium through which to travel.

_____9. This type of lens is also called a "diverging" lens.

____10. The lowest point of a wave.

Part II—Multiple Choice

For each of the following questions, choose the best answer.

11. Which of the following is not an example of a transverse wave?
   a) visible light
   b) gamma wave
   c) x-ray
   d) sound wave

12. Which characteristic of sound waves is associated with what we call "loudness"?
   a) amplitude b) frequency c) period d) wavelength

13. An incident ray of light with an angle of 35.0° from the normal strikes a plane (flat) mirror. What will be the angle of reflection, measured to the normal of the medium?
   a) 180.0° b) 90.0° c) 550.0° d) 35.0°

14. Which of the following could not be explained with the particle-theory of light?
   a) reflection b) refraction c) interference d) color

Part III—Calculations

Perform each of the following calculations.

15. The index of refraction for a particular piece of glass is 1.60. Assuming the glass is surrounded by air ($n = 1.00$) what is the critical angle of incidence for light starting inside the glass?

16. A lightbulb with a height of 10.0 cm is placed at a distance of 34.0 cm in front of a convex mirror that has a focal length of 16.0 cm. Find the distance at which the image is formed from the mirror and the height of the image.

17. A 6.50 cm tall candle is placed 14.0 cm from a concave mirror with a focal length of 6.00 cm. Find the distance at which the image is formed from the mirror and the height of the image.

Answer Key

The actual answers will be shown in brackets, followed by an explanation. If you don't understand an explanation that is given in this section, you may want to go back and review the lesson that the question came from.

Lesson 8–1 Review

1. [longitudinal wave]—Sound waves are examples of longitudinal waves.
2. [mechanical wave]—Sound, which consists of mechanical waves, can't travel through a vacuum.
3. [rarefaction]—The rarefaction of a longitudinal wave is analogous to the crest of a transverse wave.

Lesson 8–2 Review

1. [destructive interference]—Destructive interference is used in "noise reduction" technology.
2. [0.125 m]—Microwaves are electromagnetic waves, so we use the constant, C.

$$\lambda = \frac{C}{f} = \frac{3.00 \times 10^8 \text{ m/s}}{2.40 \times 10^9 \text{ Hz}} = 0.125 \text{ m}$$

3. [1.0×10^{-18} s]—Because the wave is electromagnetic, we use the constant, C.

$$T = \frac{\lambda}{C} = \frac{3.0 \times 10^{-10} \cancel{\text{m}}}{3.00 \times 10^8 \cancel{\text{m}}/\text{s}} = 1.0 \times 10^{-18} \text{ s}$$

Lesson 8–3 Review

1. [the Doppler effect]—We say the "apparent" change in frequency because the waves aren't being generated faster, they are just reaching the listener faster, due to the motion of the source.
2. [amplitude]—A larger disturbance creates a louder sound.
3. [frequency]—A person with a high-pitched voice generates waves faster than someone with a low-pitched voice.

Lesson 8–5 Review

1. [virtual image]—Unlike real images, virtual images can't be projected onto a screen.
2. [–19 cm and 2.0×10^1 cm]—Check the sign conventions for a concave mirror. All of our given values should remain positive.

   Given: $h_o = 8.5$ cm $d_o = 8.0$ cm $f = 14.0$ cm

   Find: d_i and h_i

   Solution: $d_i = \dfrac{d_o f}{d_o - f} = \dfrac{(8.0 \text{ cm})(14.0 \text{ cm})}{(8.0 \text{ cm} - 14.0 \text{ cm})} = -18.667 \text{ cm} = -19 \text{ cm}$

The negative sign for our image distance indicates that it is formed behind the mirror, so it is virtual. This makes sense, because our object was placed at a distance less than the focal length.

$$h_i = \frac{-d_i h_o}{d_o} = \frac{-(-19\text{ cm})(8.5\text{ cm})}{8.0\text{ cm}} = 20.\text{ cm} = 2.0 \times 10^1\text{ cm}$$

Our image is enlarged and upright. Again, our answer makes sense. When a concave mirror forms a virtual image, it is enlarged and upright.

3. [–5.9 cm and 1.3 cm]—According to our sign conventions for convex mirrors, we need to make the focal length negative.
   Given: h_o = 5.0 cm d_o = 22.0 cm   f = -8.0 cm
   Find: d_i and h_i

   $$\text{Solution: } d_i = \frac{d_o f}{d_o - f} = \frac{(22.0\text{ cm})(-8.0\text{ cm})}{(22.0\text{ cm}) - (-8.0\text{ cm})} = -5.9\text{ cm}$$

   Here, the negative sign indicates that the image is formed behind the mirror.

   $$h_i = \frac{-d_i h_o}{d_o} = \frac{-(-5.9\text{ cm})(5.0\text{ cm})}{22.0\text{ cm}} = 1.3\text{ cm}$$

   Our answer for image height makes sense, indicating that the image is upright and reduced in size.

Lesson 8–6 Review

1. [1.24 × 10⁸ m/s]—$v = \frac{C}{n} = \frac{3.00 \times 10^8\text{ m/s}}{2.42} = 1.24 \times 10^8\text{ m/s}$

2. [16.4°]—$\theta_2 = \sin^{-1}\left(\frac{n_1 \sin\theta_1}{n_2}\right) = \sin^{-1}\left(\frac{(1.00)(\sin 22.0°)}{1.33}\right) = 16.4°$

3. [–7.33 cm and 1.67 cm]—Checking the sign conventions for concave lenses reveals that our focal length should be given a negative sign.
   Given: h_o = 5.00 cm d_o = 22.0 cm   f = –11.0 cm
   Find: d_i and h_i

   $$\text{Solution: } d_i = \frac{d_o f}{d_o - f} = \frac{(22.0\text{ cm})(-11.0\text{ cm})}{(22.0\text{ cm}) - (-11.0\text{ cm})} = -7.33\text{ cm}$$

   As is always the case with concave lenses, the distance of the image is negative, placing the image on the same side of the lens as the object.

   $$h_i = \frac{-d_i h_o}{d_o} = \frac{-(-7.33\text{ cm})(5.00\text{ cm})}{22.0\text{ cm}} = 1.67\text{ cm}$$

The height of the image is positive, because concave lenses always form upright images. The image is also reduced.

Chapter 8 Examination

1. [f. electromagnetic wave]—Electromagnetic waves, such as radio waves and visible light, can travel through space.
2. [j. compression]—This is where the molecules get *pressed* together.
3. [c. wavelength]—The wavelength is the length of one wave cycle.
4. [i. frequency]—Remember, frequency is measured in hertz.
5. [d. amplitude]—The amplitude can be measured from the equilibrium (rest) position to either the crest or the trough.
6. [h. longitudinal wave]—A sound wave is a good example of a longitudinal wave.
7. [k. convex]—Some people remember that a concave mirror goes in, like the opening in a cave. Convex is the opposite type.
8. [e. mechanical wave]—Sound waves are mechanical waves, so they can't travel through a vacuum.
9. [l. concave]—Don't get confused by the fact that mirrors and lenses are opposites in one sense. *Concave* lenses and *convex* mirrors are *diverging*, while *convex* lenses and *concave* mirrors are *converging*.
10. [b. trough]—The trough of a wave is shaped like the troughs that horses drink out of.
11. [d. sound wave]—All electromagnetic waves, including gamma, visible light, and radio waves, are transverse waves. Sound, on the other hand, is a longitudinal wave.
12. [a. amplitude]—The greater the amplitude of a sound wave, the more the particles of the medium are displaced, the louder the sound.
13. [d. 35.0°]—The law of reflection ($\theta_i = \theta_r$) tells us that the angle of incidence is equal to the angle of reflection, both of which are measured from the normal to the medium boundary.
14. [c. interference]—The significance of Thomas Young's double-slit experiment is that he observed a property, interference, that was not associated with macroscopic particles.
15. [38.7°]—$\theta_c = \sin^{-1}\left(\frac{n_2}{n_1}\right) = \sin^{-1}\left(\frac{1.00}{1.60}\right) = 38.7^\circ$

16. [−10.9 cm and 3.21 cm]—According to our sign conventions for convex mirrors, we need to make the focal length negative.

    Given: $h_o = 10.0$ cm $\quad d_o = 34.0$ cm $\quad f = -16.0$ cm

    Find: d_i and h_i

    Solution: $$d_i = \frac{d_o f}{d_o - f} = \frac{(34.0\text{ cm})(-16.0\text{ cm})}{(34.0\text{ cm}) - (-16.0\text{ cm})} = -10.9\text{ cm}$$

    Here, the negative sign indicates that the image is formed behind the mirror.

    $$h_i = \frac{-d_i h_o}{d_o} = \frac{-(-10.9\text{ cm})(10.0\text{ cm})}{34.0\text{ cm}} = 3.21\text{ cm}$$

    As with all images formed by convex mirrors, ours is reduced and upright.

17. [10.5 cm and −4.88 cm]—Check the sign conventions for a concave mirror first. You should see that you don't need to make any of the given values negative.

    Given: $h_o = 6.50$ cm $\quad d_o = 14.0$ cm $\quad f = 6.00$ cm

    Find: d_i and h_i

    Solution: $$d_i = \frac{d_o f}{d_o - f} = \frac{(14.0\text{ cm})(6.00\text{ cm})}{(14.0\text{ cm} - 6.00\text{ cm})} = 10.5\text{ cm}$$

    $$h_i = \frac{-d_i h_o}{d_o} = \frac{-(10.5\text{ cm})(6.50\text{ cm})}{14.0\text{ cm}} = -4.88\text{ cm}$$

    The image is inverted and reduced in size.

Heat and Thermodynamics

I'm sure that you have heard of a method for starting a fire that involves rubbing two sticks together. In this process, the kinetic energy that your hands transfer to the sticks will be converted into thermal energy because of the friction between the two sticks. As more and more kinetic energy is converted to thermal energy, the temperature of the two sticks continues to rises. Eventually, the sticks may get hot enough to ignite some kindling. In this chapter we will study heat, temperature, and the law of thermal energy.

Lesson 9–1: Heat and Temperature

Heat and temperature are two quantities that can be easily confused. Imagine cooking a very large vat of chicken soup on the stove. Let's suppose you heat the soup until it is 95°C, quite hot. You take a tablespoon and scoop out a spoonful of soup to taste. As you remove the spoonful of soup from the vat, it has the same temperature (95°C) as the larger sample. Unfortunately, as you bring the soup towards your mouth to taste it, the spoon slips from you hand, pouring its contents on your bare foot. A spoonful of 95°C soup hitting your foot hurts, but not as bad as it would if you accidentally spilled the entire vat of 95°C soup on your foot. If both the spoonful and the vat full of soup have the same temperature, why would the larger sample cause more damage if it came in contact with your skin? The answer to the question lies in the difference between temperature and heat.

Temperature is defined as a measure of the average kinetic energy of the particles of a substance. The keyword in this definition is *average*. You can have a very small sample and still have a high average, just as you could have a very large sample and still have a very small average. The molecules of soup in our spoon had the same average kinetic energy as the molecules in the vat, but the vat contained many more molecules.

Let's suppose two different classes took part in a fund-raiser. Class A, containing 14 students, collected a total of \$280. Class B, consisting of 30 students, collected \$600. The *average* amount of money collected by the students in each class would be the same, despite the fact that class B raised more total money.

$$\text{Average Class A} = \frac{\$280}{14\text{ students}} = \$20/\text{student}$$

$$\text{Average Class B} = \frac{\$600}{30\text{ students}} = \$20/\text{student}$$

There is another important reason to remember that temperature represents the *average* kinetic energy of the particles of a substance. In our fund-raising example, there could have been students who raised much more or much less than the class average. One student, for example, may have raised \$120 while another raised \$2. The class average doesn't tell us how much money each specific student raised. In much the same way, the temperature of a sample doesn't tell us the kinetic energy of a specific molecule. In any sample, there will be some molecules with greater than average kinetic energy and some with less than the average kinetic energy. This helps us understand how some molecules can evaporate, going from a liquid to gas phase, in a sample of a liquid with a temperature far below its boiling point. It also helps us to understand why heat is lost from an object or sample when evaporation occurs. The evaporating molecules represent molecules of higher than average kinetic energy, and when they leave the sample, the average kinetic energy (temperature) goes down.

Temperature Conversions

The **Kelvin scale** is the SI scale for temperature, and it is based on the concept of **absolute zero**. Absolute zero is theoretically the lowest temperature that an object can reach. At this temperature, the kinetic energy of a molecule will be zero. Although the kelvin is the SI unit for temperature, you are still likely to encounter the Celsius scale. The following formulas are used to convert between Celsius and kelvin.

$$K = {}^{o}C + 273$$

$${}^{o}C = K - 273$$

Example 1

Convert 25°C to kelvin.

To convert Celsius to kelvin, simply add 273, and drop the degree (°) symbol.

Solution: 25°C = (25 + 273) = **298 K**

The Fahrenheit scale is still used in the United States, but it is considered outdated almost everywhere else. If you need to convert between Celsius and Fahrenheit, use the following formulas.

$$°F = \frac{9}{5}°C + 32 \qquad °C = \frac{5}{9}(°F - 32)$$

Example 2

Water boils at 100°C. What is the equivalent temperature on the Fahrenheit scale?

Solution: $°F = \frac{9}{5}°C + 32 = \frac{9(100)}{5} + 32 = \mathbf{212\ °F}$

Heat is a measure of the thermal energy that is transferred from one body to another. In our soup example, the vat of soup would transfer more heat to your foot than the spoonful would because the larger sample contains a greater amount of thermal energy. Like other forms of energy, heat is measured in the SI derived units called joules (J). Another unit of energy, calories (cal), is still often used to measure heat transfer. One calorie is approximately equal to 4.19 joules.

$$1 \text{ cal} = 4.19 \text{ J}$$

When calories are discussed with reference to food, they really refer to *food calories*. One food calorie is equal to 1000 calories, or 4190 J.

Lesson 9–1 Review

1. _______________ is a measure of the average kinetic energy of the particles of a substance.
2. Convert 34.0° Celsius to kelvin.
3. Convert 76° Fahrenheit to Celsius.

Lesson 9-2: Heat Transfer

As with other forms of energy, thermal energy is neither created nor destroyed, it only changes form. When an object is heated, it must get that energy from somewhere, and when an object cools, the heat must go somewhere else. Another important aspect of heat transfer, based on the second law of thermodynamics, is that heat will be transferred from a hotter object to a colder object, not the other way around.

Heat can be transferred by one or by a combination of more than one of the following methods.

Conduction involves the transfer of heat between two objects that are in contact with each other. Molecules with relatively high kinetic energy crash into molecules with relatively low kinetic energy, and transfer some of their kinetic energy is transferred in the collision. When you immerse your body in hot bathwater, the water molecules strike your body and transfer energy through conduction.

Convection involves the transfer of energy along with a transfer of molecules with relatively high kinetic energy. If you have ever been in a bathtub in which the water has become uncomfortably cold, you may have turned on the faucet to release more hot water into the tub. The heat reaches you faster than could be expected if the kinetic energy had to transfer from molecule to molecule all the way from the source to your body. Currents of hot water will travel through the tub, and hot water will make its way to you. Convection also takes place with gas molecules, as when hot air is circulated through a room.

Radiation involves the transfer of heat in the form of electromagnetic waves. Much of the heat that reaches us from the sun is in this form. Microwave ovens are good examples of devices that heat objects via radiation, as are the "hot lamps" that are used to keep food warm in many cafeterias.

So, now you know the methods of heat transfer, and you know that heat is transferred from the hotter object to the colder object, but how much heat does a hot object transfer when it comes in contact with a colder object? Given time, two objects in contact with each other will eventually reach **thermal equilibrium**, when they both have the same temperature. The temperature at which that will occur depends on a few factors, including the masses (m) and the specific heats (C) of the objects in question. You probably already have a good understanding of mass and temperature, but let's talk more about the concept of specific heat, or, as it is often called, specific heat capacity.

Imagine placing two spoons of equal mass, one made of wood and one made of metal, in a pot of boiling water for one minute, which is long enough to transfer some heat, but not enough to bring the temperature of the spoons up to the temperature of the water. When you take the spoons out, would you expect them to be at the same temperature, or would the metal spoon be hotter than the wooden one? If you guessed the latter scenario, you are right, and you probably have some understanding of the concept of **specific heat.**

The specific heat of a substance is defined as the amount of heat required to raise the temperature of one gram of a substance by one degree Celsius. Some objects will experience a greater change in temperature than others, even when the amount of heat transferred and the mass of the individual objects are the same. If you have ever gone to the beach, you probably noticed a large difference between the temperature of the sand and the water. Often, the sand is very hot, yet the water can be cool or even cold, despite the fact that they are both under the same sun. One reason for this has to do with the specific heat of each of the different substances. The specific heat of water, 4190 J/kg · °C, or 1.00 cal/g · °C, is quite high. As a result, the temperature of water changes more slowly than the temperature of many other substances.

Heat Transfer

heat transfer = mass × specific heat × change in temperature

or

$$Q = mC\Delta T$$

Example 1

How many joules of energy are required to raise the temperature of 2.50 kg of water from 11.0°C to 33.0°C?

Given: $m = 2.50\text{ kg}$ $\quad \Delta T = T_f - T_i = 33.0°\text{C} - 11.0°\text{C} = 22.0°\text{C}$

$C = 4190\text{ J/kg} \cdot °\text{C}$

Find: Q

Solution:

$$Q = mC\Delta T = (2.50\text{ kg})(4190\text{ J/kg} \cdot °\text{C})(22.0\ °\text{C}) = \mathbf{2.30 \times 10^5\ J}$$

Notice how the units cancel out in Example 1. Make sure that if you are given a problem that deals with calories, you make the appropriate conversions or select the specific heat in terms of calories.

Example 2

A 3.50 kg sample of water with an initial temperature of 17.0°C absorbs 8840 calories of heat from its surroundings. What is the final temperature of the water?

Note, the heat transferred (Q) is given in calories, so I will select the specific heat of water in calories as one of my givens. Alternatively, I could use the conversion factor, 1 cal = 4.19 J, to convert calories to joules.

Given: m = 3.50 kg $\quad T_i = 17.0^\circ \quad Q = 8840$ cal

$C = 1.00 \times 10^3$ cal/kg · °C

Find: T_f

Isolate: Remember that $\Delta T = T_f - T_i$, so I will substitute that value into our original formula: $Q = mC(T_f - T_i)$

Next, we divide both sides by mC: $\dfrac{Q}{mC} = \dfrac{\cancel{mC}(T_f - T_i)}{\cancel{mC}}$

Rearranging, we get: $T_f - T_i = \dfrac{Q}{mC}$

Now, add T_i to both sides: $T_f - T_i + T_i = \dfrac{Q}{mC} + T_i$

Giving us our working formula: $T_f = \dfrac{Q}{mC} + T_i$

Solution:

$$T_f = \frac{Q}{mC} + T_i = \frac{8840\ \cancel{cal}}{(3.50\ \cancel{kg})(1.00 \times 10^3\ \cancel{cal}/\cancel{kg} \cdot {}^\circ C)} + 17.0^\circ C$$

$$= \mathbf{19.5^\circ C}$$

Lesson 9–2 Review

1. ________________ is the amount of heat required to raise the temperature of one gram of a substance by one degree Celsius.

2. Calculate the specific heat capacity of an unknown material if it takes 33,400 J of energy to raise the temperature of a 1.3 kg sample of the substance by 14°C.

3. How much heat, in joules, would be required to raise the temperature of 45.5 kg of water by 25.0°C?

Lesson 9–3: Thermal Expansion

When you studied chemistry, you probably learned that one of the unusual properties of water is that within a certain range (0°C to 4°C) of temperatures, it gets more dense as its temperature increases. As a solid, water is less dense, which allows ice to float on liquid water. Most substances tend to expand to some extent when they are heated. This is to be expected if you understand the definition of *temperature* that we went over in our last lesson. Molecules at greater temperatures have greater kinetic energy. When fast-moving molecules crash into each other, they bounce away to a greater degree than slow-moving molecules would. Just as fast-moving bumper cars would bounce back more from a head-on collision than slow-moving ones would. This is what causes the size of objects to increase slightly as they get hotter, which is called **thermal expansion**, and the density of such materials to decrease.

For this lesson, we will only concern ourselves with **linear expansion**, which basically involves an increase in the length of an object as its temperature increases. As you can surmise from our example of water, not all materials show the same degree of expansion as their temperatures change. The degree to which an object expands as its temperature rises depends on the material from which the object is made. The formula for linear expansion can be used to calculate the change in length an object will experience, based on the original length of the object, the change in temperature of the object, and the coefficient of linear expansion for the material from which the object is made.

Linear Expansion

change in length = coefficient of linear expansion × initial length × change in temperature

or

$$\Delta L = \alpha L_i \Delta T$$

Example 1

A bar of a certain metal has an initial length of 1.000 m at 10.0°C. When the bar is heated to 90.0°C, the length of the bar is carefully measured to be 1.002 m. What is the coefficient of linear expansion for this metal?

Given: $L_i = 1.000$ m $L_f = 1.002$ m $T_i = 10.0°C$ $T_f = 90.0°C$

Find: α

Isolate:

Starting with the original formula: $\Delta L = \alpha L_i \Delta T$

We divide both sides by $L_i\Delta T$: $\frac{\Delta L}{L_i \Delta T} = \frac{\alpha \cancel{L_i} \cancel{\Delta T}}{\cancel{L_i} \cancel{\Delta T}}$

Rearranging, we get: $\alpha = \frac{\Delta L}{L_i \Delta T}$

Solution: $\alpha = \frac{\Delta L}{L_i \Delta T} = \frac{(L_f - L_i)}{L_i(T_f - T_i)} = \frac{(1.002\text{ m} - 1.000\text{ m})}{1.000\text{ m}(90.0°\text{C} - 10.0°\text{C})}$

$= \mathbf{2.5 \times 10^{-5}\ °C^{-1}}$

Example 2

Concrete has a coefficient of linear expansion of approximately $1.20 \times 10^{-5}\ °C^{-1}$. If a stretch of concrete has a length of 25.0 m at 11.0°C, what will be the length of the stretch at 40.0°C?

Given: $\alpha = 1.20 \times 10^{-5}\ °C^{-1}$ $L_i = 25.50$ m $T_i = 11.0°C$
$T_f = 40.0°C$

Find: L_f

Isolate:

Starting with our original formula: $\Delta L = \alpha L_i \Delta T$

Substitute for ΔL: $L_f - L_i = \alpha L_i \Delta T$

Now, add L_i to both sides: $\cancel{L_i} + L_f - \cancel{L_i} = L_i + \alpha L_i \Delta T$

We get our working formula: $L_f = L_i + \alpha L_i \Delta T$

$$L_f = L_i + \alpha L_i \Delta T$$

Solution: $= 25.50\ \text{m} + (1.20 \times 10^{-5}\ ^\circ\text{C}^{-1})(25.50\ \text{m})(29.0^\circ\text{C})$

$= \mathbf{25.51\ m}$

As you might imagine, when the length of a solid object expands, its width and height are likely to expand as well. Sometimes you will want to be able to calculate the change in the volume of a substance as its temperature changes.

For most solid objects, the change in the volume due to a change in temperature can be approximated with the formula:

$$\Delta V = 3\alpha V_i \Delta T$$

When you are asked to calculate the change in volume for a liquid, you will likely be given a coefficient of volume expansion, represented by the Greek letter beta (β). The coefficient will be used in the formula for volume expansion.

Volume Expansion

Change in volume = coefficient of volume expansion × initial volume × change in temperature

or

$$\Delta V = \beta V_i \Delta T$$

Example 3

Mercury has a coefficient of volume expansion (β) of $1.6 \times 10^{-4}\ ^\circ\text{C}^{-1}$. If the temperature of a sample of mercury with an initial volume of $2.50 \times 10^{-2}\ \text{m}^3$ changes from 5.00°C to 34.0°C, what will be the final volume of the sample?

Given: $\beta = 1.6 \times 10^{-4}\ ^\circ\text{C}^{-1}$ $\Delta T = T_f - T_i = 34.0^\circ\text{C} - 5.00^\circ\text{C} = 29.0^\circ\text{C}$

$V_i = 2.50 \times 10^{-2}\ \text{m}^3$

Find: V_f

Solution:

$$V_f = V_i + \beta V_i \Delta T$$
$$= 2.50 \times 10^{-2}\ m^3 + (1.6 \times 10^{-4}\ °C)(2.50 \times 10^{-2}\ m^3)(29.0\ °C)$$
$$= \mathbf{2.51 \times 10^{-2}\ m^3}$$

Lesson 9–3 Review

1. Calculate the coefficient of volume expansion of a specific liquid if a pure sample has a volume of $2.343 \times 10^{-5}\ m^3$ at 2.00°C and a volume of $2.397 \times 10^{-5}\ m^3$ at 45.0°C.
2. A block of a certain material has an initial length of 9.452×10^{-1} m, at 6.00°C. When the block is heated to 64.0 °C, its length of the bar is carefully measured to be 9.478×10^{-1} m. What is the coefficient of linear expansion for this metal?
3. Gasoline has a coefficient of volume expansion (β) of $9.5 \times 10^{-4}\ °C^{-1}$. If the temperature of a sample of gasoline with an initial volume of $3.547 \times 10^{-1}\ m^3$ changes from 8.00°C to 57.0°C, what will be the final volume of the sample?

Lesson 9–4: The Gas Laws

Your study of physics may include a study of the gas laws that you should recall from chemistry. Such overlap is only natural in these sciences, as matter is a concern of both chemists and physicists. Fortunately, if you learned these gas laws well in chemistry class, then a brief review may be all that you need now.

The Kinetic Theory of Gases

The kinetic theory of gases explains the behaviors of gases in terms of their molecules. The theory assumes that:

1. All gases are made up of individual particles (atoms and/or molecules) that are in constant motion. These particles obey Newton's laws of motion, so they move in random straight lines, unless acted upon by an unbalanced force.
2. All collisions between these particles are considered perfectly elastic, so there is no net loss of momentum or kinetic energy.

As you study the gas laws, you will encounter the term *ideal gases*. An **ideal gas** is an imaginary gas with the following characteristics:

1. The particles of an ideal gas are treated as *point masses*, which means that they are treated as if they take up no space. The entire sample of the gas takes up space, but the space occupied by the individual particles is assumed to be insignificant.
2. The particles of an ideal gas are assumed to exert no attraction on each other.

The model of the ideal gas has been developed to avoid dealing with the unmanageable number of variables involved in real-life samples of gases. Working with "ideal gases" is similar to what you do when you ignore friction in projectile motion problems. Real gas samples can approach ideal conditions, especially when the temperature of the sample is high and the pressure on the sample is low, but they never truly reach the ideal. However, the calculations that we do here are still valid, as your results will closely approximate what you would experience in the real world.

The ideal gas laws are cumulatively concerned with the following characteristics of gas samples:

- Pressure (P)—The amount of force exerted per unit of area.
- Volume (V)—How much space the gas occupies.
- Temperature (T)—The average kinetic energy of the particles of the gas.
- Moles (mol)—A measure of the amount of particles in the sample, where 1 mole = 6.02×10^{23} particles.
- Molar mass (M) the mass of one mole of a substance.

Three constants can come into play during gas law calculations:

- Universal gas law constant (R)—R = 8.31 J/mol · K
- Boltzmann's constant (k_B)—$k_B = 1.38 \times 10^{-23}$ J/K
- Standard atmospheric pressure (atm)—
  1 atm = 1.0×10^5 N/m^2

Boyle's Law

First, let's review Boyle's law, which shows the relationship between the pressure and volume of an ideal gas

Boyle's Law

The pressure and volume of a gas at constant temperature are inversely proportional to each other.

$$P_1V_1 = P_2V_2$$

In physics, pressure is often measured in pascal (Pa), where 1 Pa = 1 N/m^2.

You might also encounter the unit atmosphere (atm): 1 atm = 1.013×10^5 Pa.

Example 1

A 3.50 L sample of hydrogen gas exerts a pressure of 2.00×10^5 Pa at 295 K. If the temperature remains the same, and the gas is allowed to expand to 7.00 L, what pressure will the sample of hydrogen exert?

Given: $V_1 = 3.50$ L　　$P_1 = 2.00 \times 10^5$ Pa
$V_2 = 7.00$ L

Find: P_2

Solution: $P_2 = \dfrac{P_1V_1}{V_2} = \dfrac{(2.00 \times 10^5 \text{ Pa})(3.50 \text{ L})}{7.00 \text{ L}} = \mathbf{1.00 \times 10^5\ Pa}$

Notice that the volume of the gas was doubled and the pressure exerted by the gas was halved. That is what we meant when we defined Boyle's law, stating that the pressure and volume of a gas at constant temperature are inversely proportional to each other. As you double one, you halve the other. If you divide one by five, you multiply the other by five. Many questions that you will encounter are designed to test if you understand this relationship between the pressure and volume of an ideal gas.

Example 2

What happens to the volume of an ideal gas at constant temperature if you triple the pressure on it? (Assume the temperature is constant.)

The pressure and volume are inversely proportional to each other, if you triple one, you divide the other variable by three.

Answer: The volume is reduced to 1/3 the original volume.

Charles's Law

Charles's Law

The volume of an ideal gas at constant pressure varies directly with its kelvin temperature.

$$\frac{V_1}{T_1} = \frac{V_2}{T_2}$$

The important thing to remember when working with Charles's law, is to make sure that you do your calculations in kelvin. The Celsius scale includes zero and negative values, which would result in negative values for volume. This, obviously, wouldn't make sense, as nothing could be smaller than zero! If the temperature is given to you in Celsius, convert to kelvin as we discussed earlier in this chapter.

Example 3

A sample of neon gas occupies 2.0 L at 22.0°C. How much space would this sample occupy at 68.0°C, assuming the pressure remains constant?

Convert: $T_1 = 22.0°C = (22.0 + 273) = 295$ K
$T_2 = 68.0°C = (68.0 + 273) = 341$ K

Given: $T_1 = 295$ K $T_2 = 341$ K $V_1 = 2.0$ L

Find: V_2

Solution: $V_2 = \frac{V_1T_2}{T_1} = \frac{(2.0\ \text{L})(341\ \cancel{\text{K}})}{295\ \cancel{\text{K}}} = \mathbf{2.3\ L}$

Example 4

A sample of argon gas occupies 352 cm^3 at 345 K. At what temperature would this sample of gas occupy 291 cm^3? (Assume pressure is constant.)

Given: $T_1 = 345$ K $V_1 = 352\ cm^3$ $V_2 = 291\ cm^3$

Find: T_2

Solution: $T_2 = \frac{T_1V_2}{V_1} = \frac{(345\ \text{K})(291\ \cancel{\text{cm}^3})}{352\ \cancel{\text{cm}^3}} = \mathbf{285\ K}$

Ideal Gas Law

The ideal gas law shows the relationship between the temperature, volume, and pressure of an ideal gas. There are several versions of the formula that can be derived from one another. I won't go over the derivations here, but I will go over several examples of problems.

Ideal Gas Law and Related Equations

1. $\frac{P_1V_1}{T_1} = \frac{P_2V_2}{T_2}$
2. $PV = nRT$ Where n is the number of moles and R is the universal gas law constant.
3. $PV = Nk_BT$ Where N is the number of particles and k_B is the Boltzmann's constant.
4. $v_{rms} = \sqrt{\frac{3k_BT}{m}}$ Where m is the mass of each molecule and v_{rms} is the root-mean-square speed of the particles.
5. $v_{rms} = \sqrt{\frac{3RT}{M}}$ Where M is the molar mass of the gas.

Example 5

How many moles of hydrogen gas would occupy 3.75 dm^3 at 42.0°C and 3.00 atm of pressure?

Remember, the units of the constant dictate the units that you must use when solving the problem. Once you realize that you are going to use a formula involving the universal gas constant (R), you should make the necessary conversions to match the units it comes with. Look at the constant and the equivalent units:

$R = 8.31\ J/mol \cdot K = 8.31\ dm^3 \cdot kPa/mol \cdot K$
$= 8.31\ (1 \times 10^{-3}\ m^3) \cdot Pa/mol \cdot K.$

It seems to me that, based on our original given units, it makes sense to use the second value for R (8.31 $dm^3 \cdot kPa/mol \cdot K$), as that will result in the least number of conversions.

Convert: $T = 42.0°C = (42.0 + 273) = 315\ K$

$$P = 3.00\ \cancel{atm} \times \frac{1.013 \times 10^2\ kPa}{1\ \cancel{atm}} = 3.04 \times 10^2\ kPa$$

Given: $V = 3.75\ dm^3$ $T = 315\ K$ $P = 3.04 \times 10^2\ kPa$
$R = 8.31\ dm^3 \cdot kPa/mol \cdot K$

Find: n

Solution:

$$PV = n = \frac{PV}{RT} = \frac{(3.04 \times 10^2\ \cancel{kPa})(3.75\ \cancel{dm}^3)}{(8.31\ \cancel{dm}^3 \cdot \cancel{kPa}/mol \cdot \cancel{K})(315\ \cancel{K})} = \mathbf{0.436\ moles}$$

Example 6

The molecular mass of helium is 4.003 g/mol. How fast do helium molecules in a gas sample with a temperature of 32.0°C travel?

Convert: $T = 32.0°C = (32.0 + 273) = 305\ K$

$$M = 4.003\ g/mol \times \frac{1 \times 10^{-3}\ kg/mol}{1\ g/mol} = 4.003 \times 10^{-3}\ kg/mol$$

Given: $T = 305\ K$ $R = 8.31\ J/mol \cdot K$ $M = 4.003 \times 10^{-3}\ kg/mol$

Find: v_{rms}

Solution: $v_{rms} = \sqrt{\frac{3RT}{M}} = \sqrt{\frac{3(8.31\ \text{J/mol}\cdot\text{K})(305\ \text{K})}{(4.003\times10^{-3}\ \text{kg/mol})}} = \mathbf{1380\ m/s}$

Lesson 9–4 Review

1. _______________ states that the volume of an ideal gas at constant pressure varies directly with its kelvin temperature.
2. A sample of nitrogen gas occupies 6.65 cm^3 at 24.0°C and 2.0 atm of pressure. If the gas is heated to 49.0°C and compressed to a volume of 5.00 cm^3, how much pressure would the gas exert in atmospheres?
3. How many molecules of carbon dioxide can be found in a 1.20 m^3 sample at a temperature of 25.0° and a pressure of 3.4×10^5 Pa?

Lesson 9–5: The Laws of Thermodynamics

Thermodynamics is the study of the conversion of energy between heat and other forms of energy. It is the interrelation between heat, work, and internal energy of a system.

The First Law of Thermodynamics

The first law of thermodynamics is a specific statement of the law of conservation of energy. It tells us that when heat is transferred to a system, it can increase the internal energy of a system and/or it can be used by the system to do work.

The First Law of Thermodynamics

Where Q = net heat gained by the system, ΔU = change in internal energy, and W = net work done by the system.

$$Q = \Delta U + W$$

A money analogy is often used to describe the first law of thermodynamics. If you suddenly came into some unexpected extra money (like extra heat energy), you could save it (increase internal energy) and/or spend it (use it to do work). Of course, any combination of the two is possible. You could save it all, spend it all, or save some and spend some.

This would be a good time to point out a formula for work that will come in handy when working with gases.

$$W = P\Delta V$$

Work = pressure × change in volume

Example 1

The initial internal energy of a system is 45 J. 32 J of energy are added to the system as heat, as the system does 21 J of work on its surroundings. What is the final internal energy of the system?

Given: $U_i = 45$ J $Q = 32$ J $W = 21$ J

Find: U_f

Isolate:

Starting with our original equation: $Q = \Delta U + W$

We isolate the change in internal energy: $\Delta U = Q - W$

Because $\Delta U = U_f - U_i$, we get: $U_f = U_i + (Q - W)$

Solution: $U_f = U_i + (Q - W) = 45\text{ J} + (32\text{ J} - 21\text{ J}) = \mathbf{56\text{ J}}$

Use the money analogy to check our answer. Suppose you had $45 in the bank (45.0 J) and you earned an extra $32 (Q = 32 J). You used $21 to buy something (W = 21 J). How much money would you have left? $45 + $32 – $21 = $56.

The Second Law of Thermodynamics

The second law of thermodynamics has several statements associated with it. The ones that are likely to be stressed in your physics class will be shown here.

1. **Heat will not spontaneously flow from a colder body to a warmer body.** If you go swimming in cold water, heat is transferred from your body to the water, not the other way around.

2. **The entropy (disorder) of any closed system is always increasing.** If your mother asks you why you can't keep your room clean you can reply, "Mom, even I can't violate the second law of thermodynamics!" However, if she knows her physics, she may point out that your room isn't a closed system.
3. **No cyclic process that converts heat entirely into work is possible.** Some energy is always "lost" due to heat. No heat engines are 100-percent efficient. It is impossible to construct a perpetual motion machine.

The Third Law of Thermodynamics

Absolute zero is the theoretical temperature at which the kinetic energy of the particles of a substance will reach zero. The third law of thermodynamics states that the entropy (disorder) of a substance approaches zero as its temperature approaches absolute zero. This temperature has never been reached experimentally, but scientists have cooled things to within a fraction of 1 K.

Thermodynamic Processes

There are several specific thermodynamic processes associated with the ideal gases that you should become familiar with.

- **Isothermal Process:** A process in which the temperature of the gas remains the same.
- **Isobaric Process:** A process in which the pressure of the gas remains the same.
- **Isometric Process:** A process in which the volume of the gas remains the same.
- **Adiabatic Process:** A process in which no heat is gained or lost by the system.

Lesson 9–5 Review

1. The pressure of an ideal gas is maintained at 2.45×10^5 Pa while its volume increases by 0.0345 m^3. How much work does the gas do on its surroundings?
2. A system that absorbs 340 J of heat from its surroundings does 80 J of work as it expands. Find the change in its internal energy.

Chapter 9 Examination

Part I—Matching

Match the following terms to the definitions that follow.

a. absolute zero
d. Boyle's law
g. temperature
b. kelvin
e. Charles's law
h. pressure
c. Celsius
f. convection
i. specific heat

_____1. The temperature scale based on absolute zero.

_____2. States that the pressure and volume of an ideal gas at constant temperature are inversely proportional to each other.

_____3. The amount of energy required to increase the temperature of one gram of a substance by one degree Celsius.

_____4. Theoretically, the lowest temperature that an object can reach.

_____5. The amount of force exerted per unit of area.

Part II—Multiple Choice

For each of the following questions, select the best answer.

6. What would happen to the pressure of a sample of neon gas if the volume remains constant as the temperature is increased from 30.0 K to 60.0 K?
   a) The pressure would increase by 30.0 Pa.
   b) The pressure would increase by a factor of three.
   c) The pressure would decrease by a factor of three.
   d) The pressure would decrease by 30.0 Pa.

7. In an isometric process, which of the following quantities remain constant?
   a) pressure
   c) heat
   b) temperature
   d) volume

8. The sun warming Earth is an example of which form of heat transfer?
   a) radiation
   c) conduction
   b) convection
   d) grounding

Part III—Calculations

Perform the following calculations.

9. A system absorbs 235 J of heat from its surroundings, and the surroundings does 80 J of work on the system. Find the change in its internal energy.

10. Calculate the specific heat capacity of an unknown material if it takes 16700 J of energy to raise the temperature of a 1.13 kg sample of the substance by 22°C.

11. What temperature in Fahrenheit is equivalent to 31.0°C?

12. What temperature in Celsius is equivalent to 431 K?

13. A 2.61 kg sample of water with an initial temperature of 4.00°C absorbs 6570 calories of heat from its surroundings. What is the final temperature of the water?

14. Brass has a coefficient of linear expansion of approximately 1.9×10^{-5} °C^{-1}. If a bar of brass has an initial length of 1.922 m at 3.00°C, what will be the length of the bar at 121.0°C?

15. The pressure of an ideal gas is maintained at 1.55×10^5 Pa while its volume increases by 0.1777 m^3. How much work does the gas do on its surroundings?

16. A 7.50 L sample of hydrogen gas exerts a pressure of 1.25×10^5 Pa at 299 K. If the temperature remains the same, as the gas is compressed to 2.00 L, what pressure will the sample of hydrogen exert?

17. A sample of argon gas occupies 157 cm^3 at 311 K. At what temperature would this sample of gas occupy 101 cm^3? (Assume pressure is constant.)

18. How many moles of neon gas would occupy 1.75 dm^3 at 51.0°C and 1.50 atm of pressure?

19. The molecular mass of nitrogen is 28.0 g/mol. How fast do nitrogen molecules in a gas sample with a temperature of 57.0°C travel?

Answer Key

The actual answers will be shown in brackets, followed by an explanation. If you don't understand an explanation that is given in this section, you may want to go back and review the lesson that the question came from.

Lesson 9–1 Review

1. [temperature]—Remember, the keyword in this definition is *average*.
2. [307 K]—$K = {}^\circ C + 273 = 34.0 + 273 = 307\ K$
3. [24°C]—$^\circ C = \frac{5}{9}({}^\circ F - 32) = \frac{5}{9}(76 - 32) = 24^\circ C$

Lesson 9–2 Review

1. [specific heat or specific heat capacity]—Objects with low specific heat capacities will heat up quicker.
2. [1800 J/kg · °C]—$C = \frac{Q}{m\Delta T} = \frac{33400\ J}{(1.3\ kg)(14^\circ C)} = 1800\ J/kg \cdot {}^\circ C$
3. [4.77×10^6 J]—

   $Q = mC\Delta T = (45.5\ kg)(4190\ J/kg \cdot {}^\circ C)(25.0\ {}^\circ C) = 4.77 \times 10^6\ J$

Lesson 9–3 Review

1. [$5.36 \times 10^{-4}\ {}^\circ C^{-1}$]—

   $$\beta = \frac{\Delta V}{V_i \Delta T} = \frac{(V_f - V_i)}{V_i(T_f - T_i)} = \frac{(2.397 \times 10^{-5}\ m^3) - (2.343 \times 10^{-5}\ m^3)}{2.343 \times 10^{-5}\ m^3 (45.0^\circ C - 2.00^\circ C)}$$
   $$= 5.36 \times 10^{-4}\ {}^\circ C^{-1}$$

2. [$4.74 \times 10^{-5}\ {}^\circ C^{-1}$]—

   Given: $L_i = 9.452 \times 10^{-1}$ m $L_f = 9.478 \times 10^{-1}$ m $T_i = 6.00^\circ C$
   $T_f = 64.0^\circ C$

   Find: α

   Solution: $$\alpha = \frac{\Delta L}{L_i \Delta T} = \frac{(L_f - L_i)}{L_i(T_f - T_i)} = \frac{(9.478 \times 10^{-1}\ m) - (9.452 \times 10^{-1}\ m)}{(9.452 \times 10^{-1}\ m)(64.0^\circ C - 6.0^\circ C)}$$
   $$= 4.74 \times 10^{-5}\ {}^\circ C^{-1}$$

3. [$3.7 \times 10^{-1}\ m^3$]—

Given: $\beta = 9.5 \times 10^{-4}\ °C^{-1}$ $\Delta T = T_f - T_i = 57.0°C - 8.00°C = 49.0°C$

$V_i = 3.547 \times 10^{-1}\ m^3$

Find: V_f

Solution: $V_f = V_i + \beta V_i \Delta T$

$= 3.547 \times 10^{-1}\ m^3 + (9.5 \times 10^{-4}\ °C^{-1})(3.547 \times 10^{-1}\ m^3)(49.0\ °C)$

$= 3.7 \times 10^{-1}\ m^3$

Lesson 9–4 Review

1. [Charles's law]—The formula for Charles's law is: $\frac{V_1}{T_1} = \frac{V_2}{T_2}$.

2. [3.0 atm]—Because our solution to this problem will not involve one of the constants, we can use the units that came with the problem, with the exception for the temperature, which must always be converted to kelvin.

Convert: $T_1 = 24.0°C = (24.0 + 273) = 297\ K$

$T_2 = 49.0°C = (49.0 + 273) = 332\ K$

Given: $V_1 = 6.65\ cm^3$ $T_1 = 297\ K$ $P_1 = 2.0\ atm$

$V_2 = 5.00\ cm^3$ $T_2 = 332\ K$

Find: P_2

Solution: $P_2 = \frac{P_1 V_1 T_2}{T_1 V_2} = \frac{(2.0\ atm)(6.65\ cm^3)(332\ K)}{(297\ K)(5.00\ cm^3)} = 3.0\ atm$

3. [9.92×10^{25} molecules]—When you want to solve for the number of particles in a sample, we use the formula $PV = Nk_B T$, which contains the Boltzmann constant (k_B), with a value of 1.38×10^{-23} J/K. Remember, as with our other gas law constant, keep in mind that $1\ J = (1 \times 10^{-3}\ m^3) \cdot Pa$. This means that we can use the units for volume and pressure that came with our problem. We only need to convert the temperature to kelvin.

Convert: $T = 25.0°C = (25.0 + 273) = 298\ K$

Given: $P = 3.4 \times 10^5\ Pa$ $V = 1.20\ m^3$ $T = 298\ K$ $k_B = 1.38 \times 10^{-23}\ J/K$

Find: N

Solution: $N = \frac{PV}{k_B T} = \frac{(3.4 \times 10^5\ Pa)(1.20\ m^3)}{(1.38 \times 10^{-23}\ J/K)(298\ K)} = 9.92 \times 10^{25}$ molecules

Lesson 9–5 Review

1. [8.45×10^3 J]—$W = P\Delta V = (2.45 \times 10^5\ Pa)(0.0345\ m^3) = 8.45 \times 10^3\ J$

2. [260 J]—Given: $Q = 340\text{ J}$ $W = 80\text{ J}$
Find: ΔU
Solution: $\Delta U = Q - W = 340\text{ J} - 80\text{ J} = 260\text{ J}$

Chapter 9 Examination

1. [b. kelvin]—Remember to do all gas law calculations in kelvin, so that you don't end up with negative values for quantities, such as volume, that have no negative values.
2. [d. Boyle's law]—The formula for Boyle's law shows this relationship. $P_1V_1 = P_2V_2$
3. [i. specific heat]—Specific heat is also sometimes called "specific heat capacity."
4. [a. absolute zero]—Although this temperature has never been achieved, it can be calculated.
5. [h. pressure]—Pressure is often measured in N/m^2.
6. [b. The pressure would increase by a factor of three.]—According to the ideal gas law, the pressure and temperature of the gas will vary directly, so tripling the temperature at constant volume will triple the pressure.

$$P_2 = \frac{P_1\cancel{V_1}T_2}{T_1\cancel{V_2}} = \frac{P_1T_2}{T_1}$$

7. [d. volume]—Think of "metric" and "meter."
8. [a. radiation]—All heat transferred by electromagnetic waves represents radiation.
9. [315 J]—$\Delta U = Q - W = 235\text{ J} - (-80\text{ J}) = 315\text{ J}$
10. [$670\text{ J/kg}\cdot{}^\circ\text{C}$]—$C = \frac{Q}{m\Delta T} = \frac{16700\text{ J}}{(1.13\text{ kg})(22^\circ\text{C})} = 670\text{ J/kg}\cdot{}^\circ\text{C}$
11. [88°F]—$^\circ F = \frac{9}{5}{}^\circ C + 32 = \frac{9}{5}(31.0^\circ C) + 32 = 87.8\text{ F} = 88^\circ F$
12. [158°C]—$^\circ C = K - 273 = 431 - 273 = 158^\circ C$
13. [6.52°C]—

$$T_f = \frac{Q}{mC} + T_i$$

$$= \frac{6570\text{ cal}}{(2.61\text{ kg})(1.00 \times 10^3\text{ cal/kg}\cdot{}^\circ\text{C})} + 4.00^\circ\text{C} = 6.52^\circ\text{C}$$

14. [1.9 m after rounding]—

$$\begin{aligned} L_f &= L_i + \alpha L_i \Delta T \\ &= 1.922\text{ m} + (1.9 \times 10^{-5}\ ^\circ\text{C}^{-1})(1.922\text{ m})(118.0^\circ\text{C}) \\ &= 1.926\text{ m} \end{aligned}$$

15. $[2.75 \times 10^4 \text{ J}]$— $W = P\Delta V = (1.55 \times 10^5 \text{ Pa})(0.1777 \text{ m}^3) = 2.75 \times 10^4 \text{ J}$

16. $[4.69 \times 10^5 \text{ Pa}]$— $P_2 = \dfrac{P_1 V_1}{V_2} = \dfrac{(1.25 \times 10^5 \text{ Pa})(7.50 \text{ L})}{2.00 \text{ L}} = 4.69 \times 10^5 \text{ Pa}$

17. $[2.00 \times 10^2 \text{ K}]$— $T_2 = \dfrac{T_1 V_2}{V_1} = \dfrac{(311 \text{ K})(101 \cancel{\text{cm}^3})}{157 \cancel{\text{cm}^3}} = 2.00 \times 10^2 \text{ K}$

18. [0.0991 moles]—

Convert: $T = 51.0^\circ\text{C} = (51.0 + 273) = 323 \text{ K}$

$$P = 1.50 \cancel{\text{atm}} \times \frac{1.013 \times 10^2 \text{ kPa}}{1 \cancel{\text{atm}}} = 1.52 \times 10^2 \text{ kPa}$$

Given: $V = 1.75 \text{ dm}^3$ $T = 323 \text{ K}$ $P = 1.52 \times 10^2 \text{ kPa}$

$R = 8.31 \text{ dm}^3 \cdot \text{kPa/mol} \cdot \text{K}$

Find: n

Solution: $PV = n = \dfrac{PV}{RT} = \dfrac{(1.52 \times 10^2 \cancel{\text{kPa}})(1.75 \cancel{\text{dm}^3})}{(8.31 \cancel{\text{dm}^3} \cdot \cancel{\text{kPa}}/\text{mol} \cdot \cancel{\text{K}})(323 \cancel{\text{K}})}$

$= 0.0991 \text{ moles}$

19. [542 m/s]—

Convert: $T = 57.0^\circ\text{C} = (57.0 + 273) = 330. \text{ K}$

$$M = 28.0 \text{ g/mol} \times \frac{1 \times 10^{-3} \text{ kg/mol}}{1 \text{ g/mol}} = 28.0 \times 10^{-3} \text{ kg/mol}$$

Given: $T = 330. \text{ K}$ $R = 8.31 \text{ J/mol} \cdot \text{K}$ $M = 28.0 \times 10^{-3} \text{ kg/mol}$

Find: v_{rms}

Solution: $v_{rms} = \sqrt{\dfrac{3RT}{M}} = \sqrt{\dfrac{3(8.31 \text{ J/mol} \cdot \text{K})(330. \text{ K})}{(28.0 \times 10^{-3} \text{ kg/mol})}} = 542 \text{ m/s}$

Nuclear Physics

In most chemistry books, the lessons on atomic structure are found near the beginning. Yet, in most physics books, they are found near the end. Perhaps this is because students need to understand many of the other topics in physics before they can hope to understand the complex structure of the atom. Another reason is because this chapter marks the beginning of a newer branch of study. Up until now, we have been studying mainly macroscopic objects, which are big enough to see with the naked eye, and they appear to follow the laws of Newtonian mechanics. The microscopic, subatomic particles that we study in this chapter don't appear to follow all of those laws. A new set of rules that allow for strange observations, such as mass being transformed into energy, apply here.

Lesson 10–1: Structure of the Atom

If you studied chemistry, then you probably know the basics of atomic structure. We reviewed some of the subatomic particles earlier in this book, but now we will go over the structure of the atom in a more formal way.

There are three main types of subatomic particles:

1. The **proton** is a positively charged subatomic particle normally found in the nucleus of the atom. Each proton has a charge of $+1.60 \times 10^{-19}$ C, and a mass of approximately $1.672\,65 \times 10^{-27}$ kg, or 1.007 825 atomic mass units (μ). Because protons are normally found in the nucleus of the atom, they, along with neutrons, are sometimes called **nucleons**. Protons are also classified as **hadrons**, because they are made up of smaller particles called **quarks**. The number of protons in the nucleus

of an atom is called the **atomic number** (Z), or **nuclear charge**. The number of protons in the nucleus of an atom determines its identity. For example, every hydrogen atom has only one proton in its nucleus.

2. The **neutron** is a neutrally charged subatomic particle that is also typically found in the nucleus. The neutron has a neutral charge and a mass of approximately $1.674\ 95 \times 10^{-27}$ kg, or 1.008 665 atomic mass units (μ). Neutrons and protons are sometimes collectively called nucleons because they make up the nucleus of atoms. They are also examples of hadrons because they are made up of smaller particles called quarks. Atoms of a particular element that have different numbers of neutrons in their nucleus are called isotopes. For example, the isotope of hydrogen called protium has one proton and zero neutrons in its nucleus, while the isotope of hydrogen called deuterium has one proton and one neutron in its nucleus.

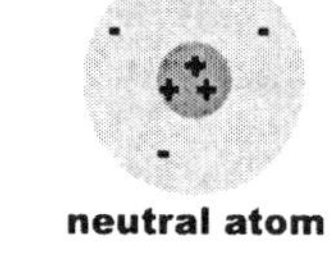

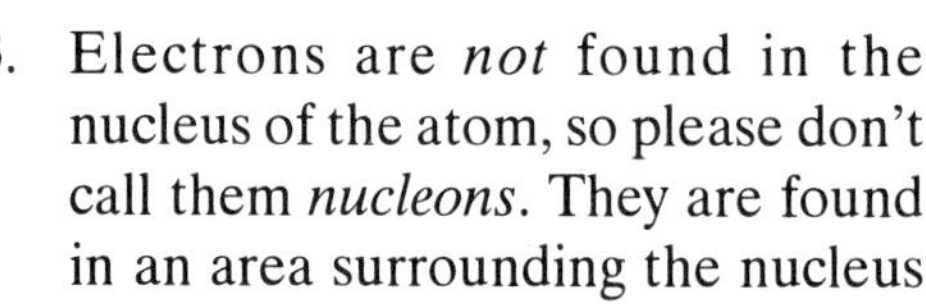

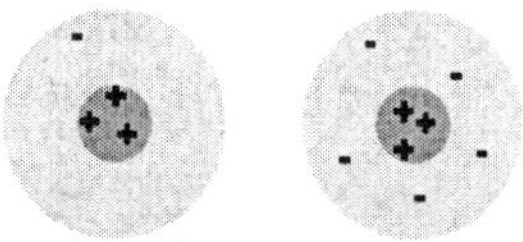

Figure 10.1

3. Electrons are *not* found in the nucleus of the atom, so please don't call them *nucleons*. They are found in an area surrounding the nucleus called the **electron cloud**. The mass of each electron is only about 9.109×10^{-31} kg, or 5.49×10^{-4} μ. Electrons are thought to be truly "elementary particles" belonging to a class called **leptons**. An atom can lose one or more electrons to become a **positive ion**, or it can gain one or more additional electrons to become a **negative ion**.

Example 1

The mass of a proton ($m_p = 1.672\ 65 \times 10^{-27}$ kg) and the mass of an electron ($m_e = 9.109 \times 10^{-31}$ kg) may not seem all that different to you, but the mass of the electron is much smaller. In fact, the mass of the electron is usually considered insignificant. So much so that the mass of the electrons are often not even taken into account when calculating the mass of a particular atom. How many electrons would you need to add together to approximate the mass of a proton?

We can solve this by dividing the mass of a proton by the mass of an electron.

Solution:

$$\text{number of electrons} = \frac{\text{mass of proton}}{\text{mass of electron}} = \frac{1.67265 \times 10^{-27}\ \text{kg}}{9.109 \times 10^{-31}\ \text{kg}}$$

$$= \mathbf{1836\ electrons}$$

So, it would take about 1836 electrons to equal the mass of one proton. This illustrates why the mass of an electron is often overlooked. This can also help you understand why the electron exhibits wave-like properties that can be measured, as we will discuss shortly.

The tiny mass of an electron can also help you understand one reason why the more predictable planetary-model has given way to the current quantum-mechanical model of the atom. In Chapter 8 we discussed the so-called photoelectric effect, where light shining on the surface of a piece of metal can cause electrons to be ejected from the metal atoms. The German physicist Werner Heisenberg pointed out that in order to locate the position of an electron, we would have to bounce photons of light off it, and that would surely change the momentum of the electron.

The Heisenberg Uncertainty Principle

It is impossible to determine accurately both the momentum and position of an electron simultaneously.

According to our current model of the atom, we can never truly know both the accurate position and momentum of an electron. We can only discuss the location of the electrons in the electron cloud in terms of probability. The likelihood of a particular electron to be found in a given area is the closest we can hope to determine.

Lesson 10–1 Review

1. A _______________ is a positively charged particle commonly found in the nucleus of the atom.

2. What part of the atom contains all of the positive charge and essentially all of the mass?

3. Where are the electrons of an atom commonly found?

Lesson 10–2: Planck's Photons

As we discussed in Chapter 8, the particle theory of light was dealt a devastating blow by the double-slit experiment carried out by Thomas Young in the beginning of the 19th century. We also learned that Albert Einstein was able to provide new evidence for the particle theory of light in 1905, when he used the idea of **photons** of light to explain the photoelectric effect. What we haven't discussed yet is the fact that Einstein applied the work of Max Planck (1858–1947) to his explanation.

Max Planck studied a particular problem involving blackbody radiation, which we don't have room to really discuss here. In 1900, Planck developed a theory and formula that treated energy as if it were made up of discrete tiny packets called quanta, or photons. Different photons represent different amounts of energy, and the energy and frequency of the photons are proportional to each other.

$$\mathbf{E \propto f}$$

Planck introduced a proportionality constant (h), which sets the two sides of the equation equal to each other.

$$\mathbf{E = hf}$$

where Planck's constant (h) has a value of 6.63×10^{-34} J × s.

It is now common practice to calculate the energy associated with the photons of a particular frequency of light, or other type of electromagnetic wave.

Example 1

Calculate the energy of a photon with a frequency of 4.5×10^{-7} Hz.

Remember, the unit we call hertz (Hz) is really a derived unit equivalent to the inverse second (s^{-1}). To make it more obvious how the units cross out, I will substitute s^{-1} for Hz in our calculation.

Given: $f = 4.50 \times 10^{-7}$ Hz $h = 6.63 \times 10^{-34}$ J · s

Find: E

Solution: $E = hf = (6.63 \times 10^{-34}\ \text{J} \cdot \cancel{\text{s}})(4.50 \times 10^{-7}\ \cancel{\text{s}}^{-1}) = \mathbf{2.98 \times 10^{-40}\ J}$

The relationship between the frequency and wavelength of an electromagnetic wave was presented in a formula in Chapter 8 $C = f\lambda$.

Solving for frequency, we get: $f = \frac{C}{\lambda}$.

If we substitute this value for frequency into Planck's equation, we get: $E = \frac{hC}{\lambda}$.

Example 2

How much energy does a photon with a wavelength of 3.50×10^{-8} m contain?

Recall, the speed of light in a vacuum (C) is 3.00×10^{8} m/s.

Given: $\lambda = 3.50 \times 10^{-8}$ m $h = 6.63 \times 10^{-34}$ J · s

$C = 3.00 \times 10^{8}$ m/s²

Find: E

Solution: $E = \frac{hC}{\lambda} = \frac{(6.63 \times 10^{-34}\ \text{J} \cdot \cancel{\text{s}})(3.00 \times 10^{8}\ \cancel{\text{m}}/\cancel{\text{s}})}{3.50 \times 10^{-8}\ \cancel{\text{m}}} = \mathbf{5.68 \times 10^{-18}\ J}$

The amount of energy represented by our answers to Examples 1 and 2 is so tiny that it is common to ask for the answers to this type of question in electron volts (eV), which we discussed in Chapter 5. Recall that one electron volt is the amount of energy required to move the charge of one electron across a potential difference of one volt. Our conversion factor from Chapter 5 is: 1 eV = 1.60×10^{-19} J. This gives us a value for Planck's constant of:

$$6.63 \times 10^{-34}\ \text{J} \cdot \text{s} \times \frac{1\ \text{eV}}{1.60 \times 10^{-19}\ \text{J}} = 4.14 \times 10^{-15}\ \text{eV} \cdot \text{s}.$$

Example 3

Calculate the wavelength of a photon of light with an energy of 2.1 eV.

Given: $C = 3.00 \times 10^{8}$ m/s² $h = 4.14 \times 10^{-15}$ eV · s E = 2.1 eV

Find: λ

Solution: $\lambda = \frac{hC}{E} = \frac{(4.14 \times 10^{-15}\ \text{eV} \cdot \text{s})(3.00 \times 10^{8}\ \text{m/s})}{2.1\ \text{eV}} = \mathbf{5.9 \times 10^{-7}\ m}$

The Momentum of a Photon

Although we often think of light as being made up of waves, as suggested by Thomas Young's double-slit experiment, other experiments have been designed to study the particle-like properties of light. One particle-like property associated with photons, momentum, can be determined with the following formula:

$$\text{momentum (p)} = \frac{\text{Planck's constant (h)}}{\text{wavelength } (\lambda)}$$

or

$$p = \frac{h}{\lambda}$$

Example 4

Calculate the momentum of a photon with a wavelength of 4.90×10^{-6} m.

Given: $\lambda = 4.90 \times 10^{-6}$ m $\qquad$ $h = 6.63 \times 10^{-34}$ J · s

Find: p

Solution: $p = \frac{h}{\lambda} = \frac{6.63 \times 10^{-34}\ \text{J} \cdot \text{s}}{4.90 \times 10^{-6}\ \text{m}} = \mathbf{1.35 \times 10^{-28}\ kg \cdot m/s}$

Wave-Particle Duality

As the particle-like properties of light and other electromagnetic waves were being established, Louis de Broglie (1892–1987) suggested that perhaps particles also had wave-like properties. It has since been verified that the wavelength of a particle can be determined with the formula:

$$\lambda = \frac{h}{m\nu}.$$

Example 5

Calculate the wavelength of an electron with a velocity of 4.5×10^7 m/s.

Given: $m = 9.109 \times 10^{-31}$ kg $v = 4.5 \times 10^7$ m/s
$h = 6.63 \times 10^{-34}$ J · s

Find: λ

Solution: $\lambda = \frac{h}{mv} = \frac{6.63 \times 10^{-34} \text{ J} \cdot \text{s}}{(9.109 \times 10^{-31} \text{ kg})(4.5 \times 10^{7} \text{ m/s})} = \mathbf{1.6 \times 10^{-11} \text{ m}}$

So, it is not just light that exhibits wave-particle duality. All things can be treated as both waves and particles. However, as you can see in the previous formula, the wavelength of a particle is inversely proportional to its mass. The masses of the objects in our macroscopic world are large enough to render their wave-like properties insignificant. Significant wave-like properties are only detectable in particles with very tiny masses.

Lesson 10–2 Review

1. Calculate the energy associated with a photon with a wavelength of 3.7×10^{-6} m.
2. Calculate the wavelength of a proton with a velocity of 2.54×10^5 m/s.
3. Calculate the momentum of a photon with a wavelength of 7.6×10^{-4} m.

Lesson 10–3: Binding Energy

Perhaps there is something about the model of the atom that has been bothering you? In Lesson 10–1 you learned that each proton has a positive charge, and most atoms have more than one proton in their nucleus. However, when we studied Chapter 5, we learned that particles with like charges repel each other. We also learned that the electrostatic force of repulsion between like charges is greatest when the objects are close together, and a particle can't get much closer to another object than when they occupy the same nucleus. How is it that multiple protons can exist in the nucleus, when they must exert a relatively strong electrostatic force of repulsion on each other?

Based on your knowledge of Newton's second law, you can assume that the answer must be that there is another strong force involved that holds the nucleus together. This force is called the **strong force**, or the **strong nuclear force**.

Strong Nuclear Force

The fundamental force of nature that exists between nucleons that holds them together against the electrostatic force of repulsion between protons.

The strong force holding the nucleus together is related to the binding energy of the atom. Would it surprise you to learn that if you were to add up all of the individual masses of all of the subatomic particles that make up the atom, they would be greater than the mass of the actual atom? You may have heard it said that something is "greater than the sum of its parts." An atom is literally less than the sum of its parts, at least if you only focus on mass. The difference between the masses of the individual particles and the total mass of the atom is called the **mass defect**.

Where did the missing mass go? Does this violate the law of conservation of mass? Einstein's answer to these questions was $E = mC^2$. One of the most famous equations in physics shows us that mass and energy are actually equivalent, and it is the sum of all of the mass and energy in a system that is conserved. Einstein combined the laws of conservation of mass and conservation of energy into the law of conservation of mass-energy.

How does this apply to what we are discussing now? Some of the mass from the individual subatomic particles, the so-called mass defect, is converted into energy, called **binding energy**, which is used to hold the nucleons together. This is the energy that is released when an atom is split.

To calculate the binding energy of a particular isotope, we must first calculate the mass defect of the atom, then we calculate the equivalent energy using the conversion factor: $1\mu = 931$ MeV.

Example 1

The nucleus of an atom of tritium ($^{3}_{1}H$) with an atomic mass of 3.016049 μ consists of two neutrons and one proton. Calculate the mass defect (Δm) and the binding energy of this isotope of hydrogen.

Note that the mass of the electron is not taken into account in these calculations.

Given: $m_p = 1.007\ 825\ \mu$ $m_n = 1.008\ 665\ \mu$ $m_{H\text{-}3} = 3.016049\ \mu$

Find: Δm and E_b

Solution:

Δm = [(number of protons)(mass of proton) + (number of neutrons)(mass of neutron)] – mass of tritium

$\Delta m = [(1)(1.007825\ \mu) + (2)(1.008665)] - 3.016049\ \mu = \mathbf{0.009106\ \mu}$

$E_b = \Delta mC^2 = (0.009106\ \mu)(931\ \text{MeV}/\mu) = \mathbf{8.48\ MeV}$

Lesson 10–3 Review

1. ______________ is the difference between the masses of the individual particles and the total mass of the atom.

2. The nucleus of an atom of deuterium (^2_1H) with an atomic mass of 2.014102 μ consists of one neutron and one proton. Calculate the mass defect (Δm) and the binding energy of this isotope of hydrogen.

Lesson 10–4: Nuclear Reactions

Radioactivity

The strong force doesn't only exist between proton pairs, it exists between pairs of neutrons and between neutrons and protons. In unstable isotopes, the strong force is unable to compensate for the total electrostatic force of repulsion between protons, and such atoms experience natural **radioactivity**. Radioactivity is the spontaneous release of energy, which is sometimes accompanied by particles, from an atom. Several major types of radioactivity have been classified.

- Alpha Decay (α): Alpha decay involves the release of an **alpha particle**, which is identical to the nucleus of a helium atom. They consist of two protons and two neutrons, and are often represented with the notation ^4_2He, which indicates a mass of four atomic mass units and an atomic charge of two.

- Beta-minus Decay (β): There are two forms of **beta decay**. Beta-minus decay, which is often simply referred to as *beta decay*, involves the release of an electron ($^{0}_{-1}e$) from the nucleus of an atom. When an electron is ejected from the nucleus of an atom, one of the neutrons is transformed into a proton, which increases the atomic number of the atom by one and changes its identity.
- Beta-plus decay (β^+) involves the release of a **positron** ($^{0}_{+1}e$) from the nucleus. A positron is the antiparticle of the electron, having the same mass but the opposite charge. When a positron is released from the nucleus of an atom, a proton is transformed into a neutron, decreasing the atomic number of the atom by one and changing its identity.
- Gamma Decay (γ): Gamma decay involves the release of high-energy photons, commonly called gamma rays. Of the three forms of radiation, gamma rays have the most penetrating power by far. Alpha particles can be blocked by clothing or paper. A thin sheet of metal can stop beta particles. Gamma rays are capable of passing through several centimeters of lead.

Nuclear Reactions

There are many different types of nuclear reactions, but they fall into two main categories: natural radioactivity and artificial radioactivity. Natural radioactivity includes all of the nuclear reactions that take place in nature. When you look at the elements on the periodic table with an atomic number of 83 or higher, you should be aware that all of the isotopes of these elements are unstable. This means that, over time, their nuclei undergo natural radioactive decay, transforming them into other elements. You may have heard of a radioactive isotope of carbon called carbon-14. Carbon-14 is naturally radioactive, undergoing beta-decay according to the following reaction:

$$^{14}_{6}C \rightarrow ^{14}_{7}N + ^{0}_{-1}e .$$

Notice that carbon (C) is being transformed into nitrogen (N) in this reaction. Also note that both the mass numbers (top numbers) and atomic numbers (bottom numbers) are conserved between each side of the equation. This will be the key to determining the missing isotope or particle in our practice problems.

Artificial radioactivity includes scientific efforts to produce previously undiscovered elements and isotopes by forcing known isotopes to combine or fuse with other particles, as shown in this reaction:

$$^{12}_{6}C + ^{244}_{96}Cm \rightarrow ^{254}_{102}No + 2^{1}_{0}n \, .$$

In case you don't recognize it, the particle on the far right side of the equation ($^{1}_{0}n$) is a neutron. You can see that it has a mass of one and an atomic number (charge) of zero. This reaction is an example of **nuclear fusion**, where two or more smaller nuclei combine to make a larger one.

Nuclear fission is the process in which a heavy nucleus splits into two or more parts. Nuclear reactors release energy using controlled nuclear fission reactions. An example of a fission reaction is shown here:

$$^{235}_{92}U + ^{1}_{0}n \rightarrow ^{90}_{38}Sr + ^{143}_{54}Xe + 3^{1}_{0}n \, .$$

Notice that there are neutrons on both sides of the equation. The neutron on the left side of the equation is used to split the atom of uranium-235. The three neutrons on the right side of the equation can then, under the proper conditions, go on to split more atoms of uranium-235. This, in turn, will release more neutrons and will set up a process called a chain reaction.

Example 1

What does X represent in the following nuclear equation?

$$^{27}_{13}Al + X \rightarrow ^{30}_{15}P + ^{1}_{0}n$$

We can tell that this reaction represents aluminum (Al) combining with a particle (X) and then releasing a neutron to form phosphorus (P). How do we determine what the particle (X) is? We must make sure that both the atomic number and mass number are conserved. That is, they show the same total on each side of the equation.

Starting with the mass number, we have 27 on the left side and a total of (30 + 1) 31 on the right side. Therefore, the particle X must have a mass number of 4.

$$^{27}_{13}Al + ^{4}X \rightarrow ^{30}_{15}P + ^{1}_{0}n$$

As for the atomic number, we have 13 on the left and a total of (15 + 0) 15 on the right. The atomic number for the missing particle must be 2, in order to make both sides equal.

$$^{27}_{13}Al + ^{4}_{2}X \rightarrow ^{30}_{15}P + ^{1}_{0}n$$

What is the identity of our missing particle? Check the periodic table and see which element has a mass of four and an atomic number of two. The answer is helium (He).

$$^{27}_{13}Al + ^{4}_{2}He \rightarrow ^{30}_{15}P + ^{1}_{0}n$$

Be aware that this is not an example of alpha decay. It is a fusion reaction, because two lighter nuclei were combined to form a heavier one.

Answer: X = $^{4}_{2}He$

Example 2

What does X represent in the following nuclear equation?

$$^{30}_{15}P + \rightarrow ^{30}_{14}Si + X$$

Whatever our mystery particle is, it has a mass of zero. We already have a mass of 30 on each side of the equation, and we need to conserve it. So, $^{30}_{15}P + \rightarrow ^{30}_{14}Si + {}^{0}X$.

We read about two particles that have a mass of zero, a positron (beta-plus) and an electron (beta-minus). The atomic number of the particle represented by X must be equal to one, because we have 15 on the left side of the equation and only 14 on the right side.

$$^{30}_{15}P + \rightarrow ^{30}_{14}Si + ^{0}_{1}X$$

The fact that the atomic number of the mystery particle is a +1 as opposed to a –1 means that it must be a positron.

$$^{30}_{15}P + \rightarrow ^{30}_{14}Si + ^{0}_{+1}e$$

So, this reaction is an example of beta-plus decay.

Answer: X = $^{0}_{+1}e$

Lesson 10–4 Review

1. ________________ is the process in which two or more smaller nuclei combine to make a larger one.

2. What does X represent in the following nuclear equation?

$$^{238}_{92}\text{U} + \rightarrow \text{X} + ^{4}_{2}\text{He}$$

3. What does X represent in the following nuclear equation?

$$^{40}_{19}\text{K} + \rightarrow ^{40}_{20}\text{Ca} + \text{X}$$

Chapter 10 Examination

Part I—Matching

Match the following terms to the definitions that follow.

a. proton	e. nucleon	i. positron
b. neutron	f. quarks	j. beta-minus particle
c. electron	g. atomic number	k. nuclear fission
d. hadron	h. photon	l. nuclear fusion

_____1. A particle (a proton or neutron) found in the nucleus of an atom.

_____2. A positively charged particle commonly found in the nucleus of an atom.

_____3. This is the antiparticle of the electron, with a mass of zero and a charge of +1.

_____4. The process in which a heavy nucleus splits into two or more parts.

_____5. The number of protons in the nucleus of an atom.

_____6. A particle (an electron) emitted from the nucleus of an atom during beta decay.

_____7. A particle made up of quarks.

_____8. The process in which two or more smaller nuclei combine to make a larger one.

_____9. A tiny packet of energy that makes up light.

____10. Fundamental particles that are constituents of protons and neutrons.

Part II—Calculations

Perform each of the following calculations.

11. Calculate the energy of photon with a frequency of 2.10×10^{-7} Hz.
12. How much energy does a photon with a wavelength of 4.12×10^{-8} m contain?
13. Calculate the momentum of a photon with a wavelength of 2.76×10^{-6} m.
14. Calculate the wavelength of an electron with a velocity of 3.15×10^{7} m/s.
15. Calculate the binding energy of an isotope with a mass defect of 0.005291 µ.

Answer Key

The actual answers will be shown in brackets, followed by an explanation. If you don't understand an explanation that is given in this section, you may want to go back and review the lesson that the question came from.

Lesson 10–1 Review

1. [proton]—A proton is similar in mass to the neutron. Its charge is equal in magnitude to that of an electron.
2. [the nucleus]—The mass of the electrons is not considered significant.
3. [the electron cloud]—The locations of individual electrons are only discussed in terms of probability.

Lesson 10–2 Review

1. [5.38×10^{-20} J]—You may need to look up Planck's constant ($h = 6.63 \times 10^{-34}$ J · s) and the speed of light in a vacuum ($C = 3.00 \times 10^{8}$ m/s).

$$E = \frac{hC}{\lambda} = \frac{(6.63 \times 10^{-34}\ \text{J} \cdot \cancel{\text{s}})(3.00 \times 10^{8}\ \cancel{\text{m}}/\cancel{\text{s}})}{3.70 \times 10^{-6}\ \cancel{\text{m}}} = 5.38 \times 10^{-20}\ \text{J}$$

2. [1.56×10^{-12} m]—You may need to look up the mass of a proton ($m_p = 1.672\,65 \times 10^{-27}$ kg) and Planck's constant ($h = 6.63 \times 10^{-34}$ J · s).

$$\lambda = \frac{h}{mv} = \frac{6.63 \times 10^{-34}\ \text{J} \cdot \text{s}}{(1.672\,65 \times 10^{-27}\ \text{kg})(2.54 \times 10^{5}\ \text{m/s})} = 1.56 \times 10^{-12}\ \text{m}$$

3. [8.7×10^{-31} kg · m/s]—$p = \frac{h}{\lambda} = \frac{6.63 \times 10^{-34}\ \text{J} \cdot \text{s}}{7.6 \times 10^{-4}\ \text{m}} = 8.7 \times 10^{-31}\ \text{kg} \cdot \text{m/s}$

Lesson 10–3 Review

1. [mass defect]—the mass defect represents the amount of matter that has been converted into energy.
2. [0.002388 μ and 2.22 MeV]—
   Given: $m_p = 1.007\,825\ \mu$ $m_n = 1.008\,665\ \mu$
   Δm = [(# of protons)(mass of proton) + (# of neutrons)(mass of neutron)] – mass of deuterium
   $\Delta m = [(1)(1.007825\ \mu) + (1)(1.008665\ \mu)] - 2.014102\ \mu = 0.002388\ \mu$
   $E_b = \Delta m C^2 = (0.002388\ \mu)(931\ \text{MeV}/\mu) = 2.22\ \text{MeV}$

Lesson 10–4 Review

1. [nuclear fusion]
2. [$^{234}_{90}\text{Th}$]—This is an example of alpha decay. You might need to look up the atomic number of our unknown element on the periodic table to discover its identity. The balanced equation for this reaction is $^{238}_{92}\text{U} + \rightarrow {}^{234}_{90}\text{Th} + {}^{4}_{2}\text{He}$.
3. [$^{0}_{-1}\text{e}$]—This is an example of beta-minus decay. The balanced equation is $^{40}_{19}\text{K} + \rightarrow {}^{40}_{20}\text{Ca} + {}^{0}_{-1}\text{e}$.

Chapter 10 Examination

1. [e. nucleon]—The number of nucleons in a nucleus is equal to the mass number of the atom.
2. [a. proton]—The number of protons in a nucleus is equal to the atomic number of the atom.
3. [i. positron]—A positron is ejected from a nucleus during beta-plus decay.
4. [k. nuclear fission]—During this process, binding energy is released.
5. [g. atomic number]—It is the atomic number of an element that gives it its identity.
6. [j. beta-minus particle]—During beta-minus decay, the atomic number of the parent isotope will increase by one.
7. [d. hadron]—***H***adrons are ***h***eavier particles, ***l***eptons are ***l***ighter particles.
8. [l. nuclear fusion]—Scientist use fusion to produce "new" isotopes.
9. [h. photon]—The energy of a particular photon can be determined using Planck's constant.

10. [f. quarks]—Quarks are thought to be sub-subatomic particles.

11. [1.39×10^{-40} J]—Given: $f = 2.10 \times 10^{-7}$ Hz $h = 6.63 \times 10^{-34}$ J · s
    Find: E

    Solution: $E = hf = (6.63 \times 10^{-34}\ \text{J} \cdot \cancel{\text{s}})(2.10 \times 10^{-7}\ \cancel{\text{s}}^{-1}) = 1.39 \times 10^{-40}\ \text{J}$

12. [4.83×10^{-18} J]—
    Given: $\lambda = 4.12 \times 10^{-8}$ m $h = 6.63 \times 10^{-34}$ J · s $C = 3.00 \times 10^{8}$ m/s^2
    Find: E

    Solution: $E = \frac{hC}{\lambda} = \frac{(6.63 \times 10^{-34}\ \text{J} \cdot \cancel{\text{s}})(3.00 \times 10^{8}\ \cancel{\text{m}}/\cancel{\text{s}})}{4.12 \times 10^{-8}\ \cancel{\text{m}}} = 4.83 \times 10^{-18}\ \text{J}$

13. [2.40×10^{-28} kg · m/s]—
    Given: $\lambda = 2.76 \times 10^{-6}$ m $h = 6.63 \times 10^{-34}$ J · s
    Find: p

    Solution: $p = \frac{h}{\lambda} = \frac{6.63 \times 10^{-34}\ \text{J} \cdot \text{s}}{2.76 \times 10^{-6}\ \text{m}} = 2.40 \times 10^{-28}\ \text{kg} \cdot \text{m/s}$

14. [2.31×10^{-11} m]—
    Given: $m = 9.109 \times 10^{-31}$ kg $v = 3.15 \times 10^{7}$ m/s $h = 6.63 \times 10^{-34}$ J · s
    Find: λ

    Solution: $\lambda = \frac{h}{mv} = \frac{6.63 \times 10^{-34}\ \text{J} \cdot \text{s}}{(9.109 \times 10^{-31}\ \text{kg})(3.15 \times 10^{7}\ \text{m/s})} = 2.31 \times 10^{-11}\ \text{m}$

15. [4.93 MeV]—$E_b = \Delta mC^2 = (0.005291\ \mu)(931\ \text{MeV}/\mu) = 4.93\ \text{MeV}$

Glossary

Absolute Zero: Theoretically, the lowest temperature that an object can reach. At this temperature, the kinetic energy of a molecule will be zero.

Acceleration: The rate at which an object's velocity changes.

Adiabatic Process: A process in which no heat is gained or lost by the system.

Alpha Decay: The process of radioactive decay involving the release of an alpha particle from the nucleus of an atom.

Alpha Particle: A particle consisting of 2 protons and 2 neutrons, with the identical composition of the nucleus of a helium atom.

Alternating Current (AC): Electric current that periodically reverses directions.

Amplitude: The maximum displacement of the particles of a medium from their rest position.

Angle of Incidence: The angle formed between the incident, or incoming, ray and the line drawn perpendicular (normal) to the boundary of the medium.

Angle of Reflection: The angle formed between the reflected, or outgoing, ray and the line drawn perpendicular (normal) to the boundary of the medium.

Angular Displacement: The change of position of a rotating body as measured by the angle through which it rotates.

Angular Velocity: The rate at which a body rotates in a particular direction. The rate of change of an object's angular displacement.

Aphelion: The point in its orbit where a planet is furthest from the sun.

Atomic Number: The number of protons in the nucleus of an atom.

Average Velocity: The total displacement of an object divided by the time elapsed during this period of time.

Beta Decay: The process of radioactive decay involving the release of a beta particle (an electron) from the nucleus of an atom, resulting in a neutron transforming into a proton.

Beta Particle: A particle (an electron) emitted from the nucleus of an atom during beta decay.

Beta-Plus Decay: The process of radioactive decay involving the release of a positron from the nucleus of an atom, resulting in a proton transforming into a neutron.

Boyle's Law: The pressure and volume of an ideal gas at constant temperature are inversely proportional to each other.

Capacitance: The ability of a capacitor to store electrical charge as its potential rises.

Capacitor: A device that stores an electric charge.

Center of Curvature: The center of the sphere of which the mirror or lens forms a part.

Centripetal Acceleration: Acceleration directed towards the center of a circular path.

Centripetal Force: A force directed towards the center of a circular path, which is responsible for keeping an object moving in circular motion.

Charles's Law: The volume of an ideal gas at constant pressure varies directly with its kelvin temperature.

Circular Motion: The motion of a body along a circular path.

Coefficient of Friction: The ratio of the force of static or kinetic friction to the normal force.

Component Vector: One of a set of two or more vectors that can be added to form a resultant vector.

Compression: An area in a material medium where the particles are more densely packed.

Concave Lens: A diverging lens.

Concave Mirror: A curved mirror with the shiny surface on the inside of the curve.

Convex Lens: A converging lens.

Convex Mirror: A curved mirror with the shiny surface on the outside of the curve.

Conduction (as the term relates to electrostatics): The process of charging an object by bringing it into direct contact with another charged object.

Constant Acceleration: A state in which the acceleration is constant. If the acceleration is nonzero, the velocity changes at a constant rate.

Constant Velocity: A state in which neither the speed nor the direction of the object changes.

Constructive Interference: The addition of two or more waves resulting in a greater net disturbance than each of the individual waves.

Contact Force: A force between two objects that are in direct contact with each other.

Conventional Current: Visualization of current flowing out of the positive terminal into the negative terminal.

Coulomb's Law: The electrostatic force between two charged objects is directly proportional to the product of their charges and inversely proportional to the square of the distance between them.

Crest: The highest point on a wave.

Critical Angle of Incidence: The minimum angle of incidence to produce total internal reflection.

Deuterium: An isotope of hydrogen with one proton and one neutron.

Diffraction: The spreading out of a wave beyond a barrier.

Diffuse Reflection: A reflection in which the reflected rays are not parallel to each other.

Direct Current (DC): Electric current that flows in one direction.

Displacement (as the term relates to motion): A vector quantity that describes the change in an object's position in terms of direction and distance.

Destructive Interference: The addition of two or more waves resulting in a smaller net disturbance than the individual waves.

Doppler Effect: The apparent change in frequency or wavelength of a wave that is perceived by an observer moving relative to the source of the waves.

Elastic Collision: A collision in which the total kinetic energy is conserved.

Elastic Potential Energy: The energy stored in an object that has been compressed or stretched.

Electric Circuit: An arrangement of components that provides one or more pathways for electric current to flow.

Electric Current: The rate at which electrically charged particles flow through a material.

Electric Field: An area of space around a charged object in which another charged object will experience a force.

Electric Field Lines: Lines drawn to represent the magnitude and direction of the force exerted by the electric field on a positive test charge, per unit of its charge.

Electric Field Strength: How much force a charged object within the field experiences at that position, per units of its charge.

Electric Potential: The electric potential energy of a charged object, divided by its charge.

Electric Power: The rate at which electrical energy is transformed.

Electric Potential Energy: The energy that a charged object has due to its location in an electric field.

Electromagnetic Induction: The production of voltage in a conducting circuit by moving it through an external magnetic field, or by changing the magnetic field.

Electromagnetic Wave: A transverse wave that is capable of traveling through a vacuum.

Electron: A negatively charge particle commonly found in the "cloud" area surrounding the nucleus of an atom.

Electron Cloud: The area surrounding the nucleus of an atom, where the electron(s) can be found.

Electron Volt: The energy required to move an electron between two points that have a difference of potential of one volt. $1\text{ eV} = 1.6 \times 10^{-19}\text{ J}$

Electroscope: A device that is used to detect electrostatic charge.

Energy: The capacity for doing work.

Field Force: A force that acts on objects without touching them.

Focal Length: The distance measured from the vertex of a mirror or lens to its focal point.

Focal Point: The point *to* which light converges, *from* which light diverges.

Force: Something that causes a change in an object's velocity.

Free Fall: The condition of an object that is falling with uniform acceleration caused by gravity.

Frequency: The number of vibrations or cycles of motion in a second, usually measured in Hertz.

Friction: A force that exists between two objects that are touching, which resists their motion relative to each other.

Gamma Decay: The radioactive decay of an atom accompanied by the release of high-energy photons.

Gravitational Potential Energy: The energy that an object possesses due to its position in a gravitational field.

Grounding: Providing a conductive path between a charged object or a circuit to the ground.

Hadron: A particle made up of quarks.

Hooke's Law: Below the elastic limit, strain is proportional to stress.

Horizontal Projectile Motion: Projectile motion in which all of the initial velocity is along the horizontal (x) axis.

Impulse: The product of a force and the time interval during which the force acts.

Induction: The process of charging an object by bringing it near a charged object and providing a pathway for charge between the neutral object and the ground.

Inelastic Collision: A collision in which the kinetic energy is not conserved.

Inertia: An object's tendency to resist changes in motion.

Instantaneous Acceleration: The change in velocity at a particular instant.

Instantaneous Velocity: The velocity of an object at a particular instant.

Interference: The overlapping of two or more wave disturbances.

Ion: An atom that has gained or lost one or more electrons.

Isobaric Process: A process in which the pressure of the gas remains the same.

Isometric Process: A process in which the volume of the gas remains the same.

Isothermal Process: A process in which the temperature of the gas remains the same.

Isotopes: Two or more forms of atoms of the same elements, containing the same number of protons but different numbers of neutrons.

Kepler's First Law of Planetary Motion: The orbits of the planets are ellipses, with the sun at one focus of the ellipse.

Kepler's Second Law of Planetary Motion: The line joining the planet to the sun sweeps out equal areas in equal times as the planet travels around the ellipse.

Kepler's Third Law of Planetary Motion: The ratio of the squares of the revolutionary periods for two planets is equal to the ratio of the cubes of their semimajor axes.

Kinematics: The study of the motion of objects without regard to the causes of motion.

Kinetic Energy: The energy that an object possesses due to its motion.

Kinetic Friction: The force of friction between two objects that are in motion relative to each other.

Law of Reflection: The angle of incidence is equal to the angle of reflection.

Lepton: An elementary particle with very small or insignificant mass.

Linear Momentum: The product of the mass and velocity of an object.

Longitudinal Wave: A wave in which the particles of the medium vibrate back and forth along the same axis in which the wave travels.

Magnetic Declination: The angle between geographic north and magnetic north.

Magnetic Domain: A microscopic region in which the magnetic fields of atoms are aligned in the same direction.

Magnetic Field: The region of space around a magnetized object in which a magnetic force can be detected.

Magnetic Field Line: Lines representing the magnitude and direction of the magnetic field.

Magnetic Field Strength: The intensity of a magnetic field.

Mass Defect: The difference between the mass of the individual nucleons that make up a nucleus and the nucleus itself.

Mass Number: The total number of protons and neutrons in the nucleus of an atom.

Mechanical Energy: The sum of the kinetic and all forms of potential energy.

Mechanical Wave: A wave that requires a medium through which to travel.

Mechanics: The branch of physics that deals with the motion of objects.

Momentum: The product of the mass and velocity of an object.

Negative Ion: An atom that has gained one or more electrons.

Net Force: The vector sum of all of the forces acting on an object.

Newtonian Mechanics: The branch of physics dealing with the motion of macroscopic objects.

Newton's First Law of Motion: In the absence of an unbalanced force, an object in motion will remain in motion, and an object at rest will remain at rest.

Newton's Law of Universal Gravitation: Two bodies attract each other with a force that is directly proportional to the product of their masses, and inversely proportional to the square of the distance between them.

Newton's Second Law of Motion: The acceleration an object experiences is directly proportional to the net force exerted upon it and inversely proportional to its mass.

Newton's Third Law of Motion: When an object exerts a force on a second object, the second object exerts an equal and opposite force on the first object.

Normal Force: A force exerted by one object on another object in a direction that is perpendicular to the surface of contact.

Nuclear Fission: The process in which a heavy nucleus splits into two or more parts.

Nuclear Fusion: The process in which two or more smaller nuclei combine to make a larger one.

Nucleon: A particle (a proton or neutron) found in the nucleus of an atom.

Parallel Circuit: A circuit with more than one path for current to flow.

Perihelion: The point in its orbit where a planet is closest to the sun.

Period: The amount of time required to complete one vibration or cycle of motion.

Photons: Tiny packets of energy.

Polarization: The process of inducing a temporary separation of charges in a neutral object by bringing it in close proximity to a charged object.

Positive Ion: An atom that has lost one or more electrons.

Positron: The antiparticle of the electron.

Potential Difference: The difference in electric potential between two points.

Power: The rate of doing work, measured in Watts.

Pressure: The amount of force exerted per unit of area.

Primary Axis: A line drawn from the vertex of a mirror or lens, through the focal point and center of curvature.

Projectile Motion: The motion of an object that experiences free fall, after being given some amount of horizontal velocity.

Protium: An isotope of hydrogen with one proton and no neutrons in its nucleus.

Proton: A positively charged particle commonly found in the nucleus of an atom.

Pulse: A short, non-repeated disturbance.

Quantum Mechanics: The branch of physics that deals with very small scales.

Quark: Fundamental particles that are constituents of protons and neutrons.

Radian: A unit of angular measurement that is equal to 57.3^{o}.

Rarefaction: An area in a material medium where the particles are less densely packed.

Real Image: An image formed by converging rays of light.

Reflection: The turning back of a wave at the boundary to the medium.

Refraction: The deflection or bending of a wave as it passes from one medium into another.

Regular Reflection: A reflection in which the reflected rays are parallel to each other.

Resistance: The opposition to the flow of electric current.

Resolution of Vectors: Breaking a vector into its components.

Rotational Equilibrium: The state of an object in which the sum of all of the clockwise and counterclockwise torques about a pivot point add up to zero.

Rotational Motion: The motion of an object that spins around an axis.

Scalar: A quantity that can be completely described with numbers and units.

Schematic Diagram: A drawing that shows the general layout of the elements of a circuit.

Semimajor Axis: Half of the longest axis of the ellipse.

Series Circuit: An electrical circuit with only one pathway for the electric current to flow.

Specific Heat: The amount of energy required to raise the temperature of 1 gram of a substance by 1 degree Celsius.

Static Electricity: A buildup of electric charge on an object.

Static Friction: The force of friction between two objects that are not in motion relative to each other.

Tangential Speed: The instantaneous linear speed of an object in circular motion along the tangent to its circular path.

Torque: The ability of an applied force to produce rotational motion.

Total Internal Reflection: The reflection that occurs when light strikes the boundary to a medium at an angle of incidence that is greater than the critical angle.

Transverse Wave: A wave in which the disturbance is perpendicular to the direction of travel.

Trough: The lowest point of a wave.

Uniform Acceleration: See *Constant Acceleration*.

Vector: A quantity having both magnitude and direction.

Velocity: Speed in a particular direction.

Vertex: The physical center of a mirror or lens.

Virtual Image: An image that appears to be formed by diverging rays of light.

Volume: The amount of space that something occupies.

Wave: A set of oscillations or vibrations that transfer energy without a net transfer of mass.

Wavelength: The length of one wave cycle. The distance measured from a point on one cycle to the same point on the next cycle.

Weight: The force of attraction between two objects due to gravity.

Work: Thc product of the displacement and the component of the force in the direction of the displacement.

Index

A

absolute zero, 292
acceleration, 19, 39-42, 47, 85-86, 103-104, 246
  average, 39
  centripetal, 159-162
  constant, 39
  velocity vs. constant, 47-48
  equations for motion with constant, 43
  instantaneous, 39, 335
  negative, 39-40
  uniform, 39, 43-46
acceleration, the law of, 85-89
action-reaction pair of forces, 90, 92
adiabatic process, 308
air resistance, 18-19
algebra, 20-22
alpha decay, 323,
alpha particle, 323
alternating current, 209
amount of substance, 14
ampere, 208
amplitude, 259
amps, 242
angle of incidence, 269
  critical, 279
angle of refraction, critical, 278-279
angular displacement, 149-150
angular velocity, 150
aphelion, 166
applied force, 81
area, cross-sectional, 212
Aristotle, 83
arrows, vector, 29
artificial radioactivity, 324-325
astronomy, 17
atmospheric conditions, 19
atom, neutral, 174
atomic number, 316
atomic structure, 315-317
average acceleration, 39
average speed, 35
average velocity, 35
axis, primary, 270
axis, semimajor, 167

B

batteries, 225
beta-minus decay, 324
beta-plus decay, 324
binding energy, 321-323
Boyle's law, 302
Brahe, Tycho, 165
Broglie, Louis de, 320

C

calories, 293
calories, food, 293

capacitance, 195-200
cells, 226
Celsius scale, 292
center of curvature, 270
centripetal acceleration, 159-162
change in electric potential energy and work, relationship between, 191
change in electric potential energy, formula for, 189
change in position, 28
characteristics of waves, 258-265
charge, nuclear, 316
charge, static, 173
charged objects, 176
charges,
  electric, 173
  elementary, 174
  excess, 181
  point, 185
charging objects, methods for, 180-184
Charles's law, 303
chemistry, 13, 17, 300
circle conversions, 148
circle, diameter of a, 1501-151
circuit diagrams, interpreting, 221
circuit, parallel, 218-219
circuit, series, 217-218
circuits, 209, 217-226
circular motion, 156-158
classical mechanics, 27
cobalt, 234
coefficient, 15, 17-18
  of friction, 94-100
  of kinetic friction, 94, 105
  of static friction, 94
collision, elastic, 135-137
collision, inelastic, 137-138
complete destructive interference, 264
component vector, 29-30
compressions, 258
concave lens, 280
concave mirror, 269-270
concave mirrors, images formed by, 270-271
conditions, atmospheric, 19
conduction, 183, 294
conductors, 181, 211
conservation of energy, 125-129
conservation of momentum, law of, 134-139
constant acceleration, 39, 43-46
constant acceleration, constant velocity vs., 47-48
constant acceleration, equations for motion with, 43
constant velocity, 35
constant velocity, constant acceleration vs., 47-48
constant, proportionality, 121-122
constant, spring, 121-122
constant, universal gravitational, 162
constants in gas law calculations, 301
contact forces, 80
continuous wave, 256
convection, 294
conventional current, 209
converging lens, 280
conversions, temperature, 292-293
convex lens, 280
convex mirror, 269-270
convex mirrors, images formed by, 271
Copernicus, 166
cosine, 23, 67
cosine, inverse, 25
cost of electric power, the, 215-217
coulomb, 174, 185
Coulomb's law, 176, 177
counting numbers, 15
crest, 256, 258
critical angle of incidence, 279
critical angle of refraction, 278-279
cross-sectional area, 212
current, 217-219
  alternating, 209
  conventional, 209

direct, 209
electrical, 207-210
water, 207-208
current-carrying wires, magnetic fields around, 236
curvature, center of, 270
curved mirrors, equations for, 271
cycle, 259

D

deceleration, 39-40
decibel scale, 266
decimal points, zeros and, 15
declination, magnetic, 235
derived units, 14
destructive interference, 264
diagram, free-body, 80-81, 101
diagram, ray, 281
diagrams, interpreting circuit, 221
diagrams, schematic, 217
difference, definition of potential, 337
difference, potential, 192-195
diffraction, 264, 267-268
diffuse reflection, 268, 269
digits, significant, 15-17
direct current, 209
displacement, 19, 27-34, 35, 116, 119, 157
displacement, angular, 149-150
displacement, distance and, 27-34
dissipative forces, 125
distance, displacement and, 27-34
diverging lenses, 280
domains, magnetic, 234
Doppler effect, 266-267
duality, wave-particle, 320

E

Earth, magnetic field of the, 235
echolocation, 266
Einstein, Albert, 255, 268, 318
elastic collision, 135-137
elastic potential energy, 121-124
electric charge, 14, 173
electric current, 14
electric field intensity, 184-185
formula for, 184
electric field strength, 184-185
electric fields, 184-187
electric forces, 176-180
electric potential energy, 187-192
formula for change in, 189
electric potential, 186, 192
formula for, 192
electric power, 213-217
formulas for, 214-215
the cost of, 215-217
electrical circuit, 209
electrical current, 207-210
electricity, 181
static, 173
electromagnetic induction, 245-249
electromagnetic wave, 256
types of, 257
electron cloud, 316
electron volt, 193-194
electrons, 174, 179, 181, 316
electroscope, 181-182
electrostatic force, 178-179, 180
electrostatic force of repulsion, 79
electrostatics, the basic law of, 173
elementary charges, 174
energy, 14, 115, 119-125
binding, 321-323
conservation of, 125-129
elastic potential, 121-124
electric potential, 187-192
formula for change in electric potential, 189
gravitational potential, 120-121, 125, 188
kinetic, 124-125, 138, 291-293
mechanical, 125-126, 128
potential, 196
equations for curved mirrors, 271
equations for free fall, 47
equations for motion with constant acceleration, 43

equations for thin lenses, 281
equilibrium, rotational, 153-156
equilibrium, thermal, 294
ether, 257
excess charges, 181
expansion, linear, 297-300
expansion, thermal, 297-300
exponent, 17-18

F

falling freely, 47-53
Faraday, Michael, 246
Faraday's law of magnetic induction, 246, 248-249
farads, 197
Fahrenheit scale, 293
ferromagnetic, 234
field forces, 80, 163
field lines, 186-187
field strength, magnetic, 240-245
fields, electric, 184-187
fields, magnetic, 233
first law of thermodynamics, 306-307
fission, nuclear, 325
flashlight, 207, 209-210
flux, magnetic, 246
focal length, 270
focal point, 270
food calories, 293
force vectors, 82
force, 14, 79, 80, 86, 159
  applied, 81
  centripetal, 159-162
  direction of the, 242
  electrostatic, 178-179, 180
  field, 163
  net, 80
  normal, 81
  nuclear, 79
  strong, 322
  strong nuclear, 322
  unbalanced, 84
forces, 79-83
  action-reaction pair of, 90, 92
  contact, 80
  dissipative, 125
  electric, 176-180
  field, 80
formulas, 19-20
free fall, 47-53
  equations for, 47
free space, permittivity of, 199
free-body diagram, 80-81, 101
frequency, 14, 151, 259, 266
friction, 19, 81, 84, 125, 134
  coefficient of kinetic, 94, 105
  coefficient of static, 94
  coefficient of, 94-100
  kinetic, 94-95
  static, 94-95
functions, trigonometry, 23, 67
fusion, nuclear, 325

G

Galilei, Galileo, 83-84
gamma decay, 324
gamma rays, 324
gas law calculations, constants in, 301
gas law, ideal, 204
gas laws, the, 300-306
gases, ideal, 301
gases, kinetic theory of, 300-301
graph, position vs. time, 53-55
graph, velocity vs. time, 56
graphing motion, 53-61
gravitation, Newton's universal law of, 177
gravitational constant, universal, 162
gravitational potential energy, 120-121, 125, 188
gravity, 47, 79, 89, 162, 178, 184
gravity, Newton's law of, 162-165
grounding, 183

H

hadrons, 315
heat, 291-314

heat flow, 307
heat transfer, 293, 294-297
heat, specific, 295
heat, temperature vs., 291-293
Heisenberg uncertainty principle, 317
Heisenberg, Werner, 317
Henry, Joseph, 246
hertz, 151
Hooke, Robert, 255
horizontal projectile motion, 61-62
horizontal surface, motion along a, 94-100
horsepower, 131
humid air, 181
Huygens, Christian, 255
hypotenuse, 23

I

ideal gas law, 304
ideal gases, 301
images formed by concave mirrors, 270-271
images formed by convex mirrors, 271
images, real, 270
images, virtual, 270
impulse, 139-141
impulse-momentum theorem, 139
incidence, 269
incidence, critical angle of, 279
incident rays, 268, 280-281
inclined plane, 100
inclined surface, motion on an, 100-106
index of refraction, 275
induction, 183
    electromagnetic, 245-249
inelastic collisions, 137-138
inertia, 83, 85-89
instantaneous acceleration, 39
instantaneous speed, 35
instantaneous velocity, 35
insulators, 181
intensity, electric field, 184-185
intensity level, 265
interference, 263-264, 268
    complete destructive, 264
    constructive, 264
    destructive, 264
internal reflection, total, 278
International System of Measurements (SI), 14
interpreting circuit diagrams, 221
inverse cosine, 25
inverse sine, 25
inverse tangent, 25
ions, 174, 175-176
iron, 234
irregular reflection, 268
isobaric process, 308
isometric process, 308
isothermal process, 308

J

joules, 129, 293

K

Kelvin scale, 292
Kepler, Johannes, 165
Kepler's laws, 165-167
kilowatts, 215
kinematics, 27-78
kinetic energy, 124-125, 138, 292-293
kinetic friction, 94-95
kinetic friction, coefficient of, 94, 105
kinetic theory of gases, 300-301

L

law of acceleration, the, 85-89
law of conservation of momentum, 134-139
law of electrostatics, the basic, 173
law of gravity, Newton's, 162-165
law of magnetic poles, 234
law of motion,
    Newton's first, 83, 152
    Newton's second, 85-89, 98, 101, 103, 155, 246
    Newton's third, 90-94

law of reflection, 269
law of refraction, 276-277
law, ideal gas, 304
laws of motion, 79-113
laws of thermodynamics, 306-308
laws, Kepler's, 165-167
laws, the gas, 300-306
length, 14, 212
lenses, 275-284
  concave, 280
  converging, 280
  convex, 280
  diverging, 280
  equations for thin, 281
  sign conventions for, 282
  types of, 280
Lenz's law, 249
leptons, 316
light, 267-268
  wave-theory of, 257
linear expansion, 297-300
linear momentum, 132-139
  formula for, 132
linear speed, formula for, 157
longitudinal wave, 258, 259
luminous intensity, 14

M

Magnesia, 233
magnetic declination, 235
magnetic domain, 234
magnetic field of the Earth, 235
magnetic field strength, 240-245
magnetic fields, 233
  around current-carrying wires, 236
  around solenoids, 239
magnetic flux, 246
magnetic poles, 234-235
  law of, 234
magnetism, 233-254
magnets, 173-174, 233-236
mass, 14
mass defect, 322
material composition, 212
material medium, 256
materials, nonohmic, 211
materials, ohmic, 211
mathematics, 19
matter, 179
mechanical energy, 125-126, 128
mechanical wave, 256
mechanics, 27
  classical, 27
  Newtonian, 27
  quantum, 27
metals, 181
methods for charging objects, 180-184
mirror, concave, 269-270
mirror, convex, 269-270
mirrors, equations for curved, 271
mirrors, images formed by concave, 270-271
mirrors, images formed by convex, 271
mirrors, reflection and, 268-275
mirrors, sign conventions for, 272
mirrors, types of, 269-270
missing angle of triangles, 24-25
missing side of triangles, 23
momentum, 115, 132-133
  formula for linear, 132
  formula for, 320
  law of conservation of, 134-139
  linear, 132-139
  of a photon, 320
motion, 34
motion along a horizontal surface, 94-100
motion of satellites, 165-167
motion on an inclined surface, 100-106
motion problem, two-dimensional, 62
motion with constant acceleration,
  equations for, 43
motion,
  Aristotle and, 83
  circular, 156-158
  graphing, 53-61
  horizontal projectile, 61-62
  laws of, 79-113

Newton's first law of, 83, 152
Newton's second law of, 85-89, 98, 101, 103, 155, 246
Newton's third law of, 90-94
parabolic, 66-68
projectile, 61-70
rotational, 147-152
translational, 147
multiplication, rounding and, 16

N

natural radioactivity, 324-325
negative acceleration, 39-40
negative ion, 174, 316
net force, 80
net torque, 153
neutral atom, 174
neutrons, 316
Newton, Sir Isaac, 83-85, 255
Newton's first law of motion, 83, 152
Newton's law of gravity, 162-165
Newton's second law of motion, 85-89, 98, 101, 103, 155, 246
Newton's third law of motion, 90-94
Newton's universal law of gravitation, 177
Newtonian mechanics, 27
newtons, 86, 185, 242
nickel, 234
Nobel Prize for Physics, 268
nonohmic materials, 211
normal force, 81
nuclear charge, 316
nuclear fission, 325
nuclear force, 79
strong, 322
nuclear fusion, 325
nuclear physics, 315-330
nuclear reactions, 323 -327
types of, 324
nucleon, definition of, 337
nucleons, 315, 316
number of significant digits, determining the, 15-16
number, atomic, 316
numbers, counting, 15
numbers, use of, 13-15

O

objects, charged, 176
objects, methods for charging, 180-184
Oersted, Hans Christian, 236, 245
Ohm's law, 211
ohmic materials, 211
ohms, 211, 213
opposites, 173

P

parabolic motion, 66-68
parallel circuit, 218-219
parallel circuits, comparing series and, 219
particle theory, 255
parts of scientific notation, 17
perihelion, 166
period, 151, 157, 259
periodic wave, 256
permittivity of free space, 199
photon, momentum of a, 320
photons, 318-321
physics before Galileo, 83
physics examination, AP, 12
Physics, Nobel Prize for, 268
physics, nuclear, 315-330
physics, the language of, 19
Planck, Max, 318
plane, inclined, 100
planets, 166
plastic, 181
point charges, 185
point masses, 301
polarization, 182-183
poles, law of magnetic, 234
poles, magnetic, 234-235
position vs. time graph, 53-55
position, change in, 28
positive ion, 174, 316

positron, 324
potential difference, 192-195
  formula for, 193
potential energy, 196
potential, electric, 192
power companies, 215
power ratings, 215
power, 115, 129-132
  electric, 213-217
  formula for, 130
  formulas for electric, 214-215
  the cost of electric, 215-217
primary axis, 270
principle of waves, the superposition of, 263
principle, Heisenberg uncertainty, 317
problem, relative motion, 37-38
problem, two-dimensional motion, 62
processes, thermodynamic, 308
projectile motion, 61-70
properties of waves, 258-265
proportionality constant, 121-122
protons, 174, 315
pulse, 255
push, 80
Pythagorean theorem, 24, 32-33

Q

quantities, scalar, 28-30
quantities, vector, 28-30, 39, 80, 86
quantum mechanics, 27
quarks, 315

R

radar, 266
radians, 148-149
radiation, 294
radio waves, 262
radioactivity, 323-324
  artificial, 324-325
  natural, 324-325
radius, 157
ramp, 100
rarefactions, 258
ratings, power, 215
ray diagram, 281
rays, incident, 268, 280-281
rays, reflected, 268
reactions, nuclear, 323-327
  types of, 324
real gas samples, 301
real images, 270
reference point, 28
reflected rays, 268
reflection, 264, 267
  diffuse, 268, 269
  irregular, 268
  law of, 269
  mirrors and, 268-275
  regular, 268
  total internal, 278
refraction, 264, 267, 275-284
  critical angle of, 278-279
  index of, 275
  law of, 276-277
regular reflection, 268
relative motion problem, 37-38
repulsion, electrostatic force of, 79
resistance, 210-213
  air, 18-19
  factors that affect, 212
  formula for, 210
resistivity, 212-213
resistors, 213
resolution of vectors, 66
resultant vector, 29-30
right triangles, 22-25
right-hand rule, 236-239, 243
rotational equilibrium, 153-156
rotational motion, 147-152
rounding, 15, 16-17
rubber, 181
rules of work, 189-190

S

samples, real gas, 301
satellites, the motion of, 165-167

scalar quantities, 28-30
scalars, 28
scale, decibel, 266
schematic diagrams, 217
scientific notation, 17-18, 88
second law of thermodynamics, 307-308
semiconductors, 211
semimajor axis, 167
series circuit, 217-218
series circuits, comparing parallel and, 219
SI base units, 14
sign conventions, 18
  for lenses, 282
  for mirrors, 272
significant digits, 15-17
sine, 23, 67
  inverse, 25
Snell's law, 276
solenoids, magnetic fields around, 239
sound waves, 265-267
space, permittivity of free, 199
specific heat, 295
speed of a wave, 260-261
speed, 33-39, 133
  angular, 150
  average, 35
  formula for linear, 157
  formula for tangential, 158
  instantaneous, 35
  tangential, 157-159
sprint constant, 121-122
static charge, 173
static cling, 173-174, 181
static electricity, 173
static friction, 94-95
  coefficient of, 94
straight wires, applying the right-hand rule to, 242
strength, electric field, 184-185
strength, magnetic field, 240-245
strong force, 322
strong nuclear force, 322
structure of the atom, 315-317
substance, amount of, 14
subtraction, rounding and, 17
superposition principle of waves, the, 263
surface, motion along a horizontal, 94-100
surface, motion on an inclined, 100-106
symbols, 19-20

T

tail-to-tip method, 30, 32
tangent, 23, 67
tangent, inverse, 25
tangential speed, 157-159
  formula for, 158
temperature conversions, 292-293
temperature, 14, 212, 291
temperature, heat vs., 291-293
tesla, 242
theorem, impulse-momentum, 139
theorem, Pythagorean, 24
theory of gases, kinetic, 300-301
theory, particle, 255
theory, wave, 255
thermal equilibrium, 294
thermal expansion, 297-300
thermodynamic processes, 308
thermodynamics,
  first law of, 306-307
  laws of, 306-308
  second law of, 307-308
  third law of, 308
third law of thermodynamics, 308
time, 14, 157
tip-to-tail method, 29, 32
torque, 152-156
  formula for, 153
  net, 153
total internal reflection, 278
translational motion, 147
transverse wave, 257, 258
triangles, missing angle of, 24-25
triangles, missing side of, 23-24
triangles, right, 22-25

trigonometry, 22-25
functions of, 23, 67
trough, 256, 258
two-dimensional motion problem, 62

U

unbalanced force, 84
uniform acceleration, 39
units, derived, 14
units, use of, 13-15
universal gravitation, definition of Newton's law of, 337
universal gravitational constant, 162
use of numbers, 13-15
use of units, 13-15

V

vacuum permittivity, 199
vector arrows, 29
vector quantities, 28-30, 39, 80, 86
vector sum, 83
vector, component, 29-30
vector, definition of, 339
vector, resultant, 29-30
vectors, definition of resolution of, 338
vectors, force, 82
vectors, resolution of, 66
velocity vs. time graph, 56
velocity, 34-39, 79, 133
angular, 150
average, 35
constant, 35
constant, constant acceleration vs., 47-48
instantaneous, 35
virtual images, 270
volt, electron, 193-194
voltage, 192, 194

W

water current, 207-208
watts, 129, 213
wave theory, 255
wave train, 256
wave, continuous, 256
wave,
electromagnetic, 256
longitudinal, 258, 259
mechanical, 256
periodic, 256
speed of a, 260-261
transverse, 257, 258
waveform, 259
wavelength, 259
wave-particle duality, 320
waves, 255-267
characteristics of, 258-265
properties of, 258-265
radio, 262
sound, 265-267
the superposition principle of, 263
waves, types of, 255-258
electromagnetic, 257
wave-theory of light, 257
weber, 246
weight, 81, 87-88
formula for calculating, 88
weightlessness, 84
wires, magnetic fields around current-carrying, 236
work, 115-119, 124
rules of, 189-190
work and change in electric potential energy, relationship between, 191

Y

Young, Thomas, 255

Z

zero, absolute, 292

About the Author

Greg Curran has been teaching science for more than 15 years. He is currently teaching chemistry and physics at Fordham Preparatory School, where he serves as the chair of the Science department. He also teaches online science classes through the University of Wisconsin, Cambridge College, the Academy of Teaching Excellence, and the JASON Academy. He is the author of *Homework Helpers: Chemistry* as well as the popular Science Help Online Website, which is designed to help students who are learning chemistry. The site can be found at *http://fordhamprep.org/gcurran*. Greg has a MS in education from Fordham University and a BA in biology from SUNY Purchase. He lives in Garrison, New York, with his wife, Rosemarie, and children James, Amanda, and Jessica.